D0206196

PHILOSOPHY OF SCIENCE AND
HISTORICAL ENQUIRY

Philosophy of Science and Historical Enquiry

JOHN LOSEE

CLARENDON PRESS . OXFORD

1987

Oxford University Press, Walton Street, Oxford OX2 6DP

Oxford New York Toronto
Delhi Bombay Calcutta Madras Karachi
Petaling Jaya Singapore Hong Kong Tokyo
Nairobi Dar es Salaam Cape Town
Melbourne Auckland
and associated companies in
Beirut Berlin Ibadan Nicosia

Oxford is a trade mark of Oxford University Press

Published in the United States
by Oxford University Press, New York

© John Losee 1987

All rights reserved. No part of this publication may be reproduced,
stored in a retrieval system, or transmitted, in any form or by any means,
electronic, mechanical, photocopying, recording, or otherwise, without
the prior permission of Oxford University Press

British Library Cataloguing in Publication Data
Losee, John
Philosophy of science and historical
enquiry.
1. Science—Philosophy
I. Title
501 Q175
ISBN 0–19–824946–2

Library of Congress Cataloging-in-Publication Data
Losee, John.
Philosophy of science and historical enquiry.
Includes indexes.
1. Science—Philosophy. 2. Science—History.
I. Title.
Q175.L664 1987 501 86–23550
ISBN 0–19–824946–2

Printed and bound in Great Britain
by Billing & Sons Limited, Worcester.

Preface

HISTORIANS of science regularly judge whether particular decisions of scientists conform to the evaluative standards of the time. Philosophers of science may render such judgements as well. However, many philosophers of science also accept a prescriptive role for their discipline. They are not content merely to collect cases in which scientific practice does, or does not, conform to selected standards. Rather they seek to formulate and recommend criteria that *ought* to govern evaluative practice. Among these criteria are criteria to gauge the evidential support provided for hypotheses by observation reports, criteria to rank competing theories, and criteria to assess the cogency of diverse types of explanation. Philosophers of science may disagree about the content of proper evaluative practice, but an underlying assumption of prescriptive philosophy of science is that conformity to evaluative standards is a necessary condition for the creation of 'good science'. Prescriptive philosophy of science thus sanctions a distinction between correct and incorrect evaluative practice. To work within its tradition is to accept the possibility of adverse judgements about present evaluative practice in science.

It is a task for the historian of science to record the evaluative standards that are explicit or implicit within scientific practice in diverse contexts. The historian of science may also seek to catalogue cases in which scientific practice does or does not conform to accepted standards. From the standpoint of prescriptive philosophy of science, success in this enterprise is at best a preparatory stage for what is really important, namely the formulation of standards, application of which constitutes good evaluative practice in science. The appropriateness of this view of the relationship between philosophy of science and history of science is the principal concern of this book.

Acknowledgements

Some of the material on the views of Whewell and Mill, Shapere, and Toulmin has previously appeared in print. I am grateful to the editors of *Studies in History and Philosophy of Science* for permission to reproduce in Chapters 6 and 7 material from 'Whewell and Mill on the Relation Between Philosophy of Science and History of Science', 14 (1983), 113–26, and 'Limitations of an Evolutionist Philosophy of Science', 8 (1977), 349–52. I am grateful as well to the editors of *The British Journal for the Philosophy of Science* for permission to reproduce in Chapter 5 material from 'Shapere's Project for a Nonpresuppositionist Philosophy of Science', 37 (1986), 233–9. I am indebted to Mrs. Hilda Cooper for invaluable assistance in the preparation of the manuscript.

Contents

List of Tables

List of Figures

I

Philosophy of Science and History of Science: The Principal Alternatives

PHILOSOPHY of science (*PS*) and history of science (*HS*)[1] are second-order interpretations of a first-order subject matter. If there were no scientific practice, then there would be no *PS* and no *HS*. However, I take it to be incontrovertible that some human activities do qualify as the practice of science. *PS* and *HS* are interpretations of this practice, and there are several possibilities for a relationship between them.

One possibility is that *PS* and *HS* are mutually exclusive interpretations of science. On this view, *PS* and *HS* share a first-order subject matter, but the practice of *PS* precludes historical considerations and the practice of *HS* precludes philosophical considerations.

A second possibility is that one of the disciplines is dependent on the other. *PS* may be dependent on *HS*, or vice versa, in either a strong sense or a weak sense. In the strong sense, no *PS* can be undertaken without historical enquiry (or vice versa). In the weak sense, there are at least some aspects of *PS* that require historical enquiry (or vice versa).

A third possibility is that the two disciplines are interdependent such that the practice of *PS* requires historical enquiry *and* the practice of *HS* requires philosophical analysis. *PS* and *HS* may be interdependent in either the strong sense or the weak sense mentioned above.

A fourth possibility is that there is a *de facto* overlap of disciplines. It might be the case that, although neither discipline is dependent on the other, certain of the conclusions reached within *PS* and *HS* are the same.

---

[1] '*PS*' and '*HS*' are taken to stand for 'philosophy of science' and 'history of science' respectively. The abbreviation '*HS*' retains the ambiguity of 'history of science'. '*HS*' may refer to a sequence of human actions, or it may refer to a purposeful interpretation of the available records about past actions. I shall use '*HS*' in the latter sense throughout.

Finally, *PS* may be subsumed under *HS*.[2] On this view, *PS*, like the history of botany and the history of physics, would be a part of *HS*. *PS* would be that part of *HS* within which narratives are developed that reconstruct the evaluative practices of scientists. A *PS* subsumed under *HS* would have no distinctive methodology of its own. Rather, its methodological standards would be those standards appropriate to *HS*.

---

[2] I assume that the subsumption of *HS* under *PS* is not a plausible position on the relationship between the two disciplines.

2

Are Philosophy of Science and History of Science Mutually Exclusive Disciplines?

I KUHN'S GESTALT ANALOGY

IN an essay published in 1977,[1] Thomas Kuhn raised the possibility that *PS* and *HS* are mutually exclusive disciplines. He suggested that *PS* and *HS* may be related to one another as *Gestalt* perspectives of a visually ambiguous figure. When historian and philosopher confront the same episode, the one sees a 'duck' and the other sees a 'rabbit'. A given enquirer may shift back and forth between the two perspectives, but 'no amount of ocular exercise and strain will educe a duck-rabbit'.[2] The philosopher of science may say to the historian of science, 'that "duck" you have described also may be viewed as a "rabbit"'. But the philosopher of science practises his craft only by bracketing out distinctively historical concerns, and vice versa.

The scholar who appreciates this contrast between the two disciplines may choose to practise alternately *PS* and *HS*. Kuhn noted that he himself had pursued this course of action.

But why, apart from personal interests, should a scholar practise such alternation? What does he gain as historian of science, for example, by practising *PS*? Of course, the historian of science should be knowledgeable about the evaluative standards and practices of the era whose history he seeks to reconstruct, but of what relevance to the historian is current philosophy of science? Indeed, preoccupation with contemporary philosophical problems may be antithetical to historical reconstruction. For instance, a historian of early modern science may be less likely to render anachronistic interpretations if he knows nothing at all about current debates in *PS*.

---

[1] Thomas S. Kuhn, 'The Relations Between the History and the Philosophy of Science' in *The Essential Tension* (Chicago: University of Chicago Press, 1977), 3–20.

[2] Ibid., 6.

Kuhn exacerbated the problem by contrasting the 'mental sets' of historian and philosopher. Whereas the historian seeks to develop a narrative which 'renders plausible and comprehensible the events it describes', the philosopher seeks 'to discover and state what is true at all times and places rather than to impart understanding of what occurred at a particular time and place'.[3] He posed the following question: 'given the deep and consequential differences between the two enterprises, what can they have to say to each other?'[4]

'Not much,' replied Ronald Giere, J. J. C. Smart, and Paul Feyerabend.[5] Giere has suggested that the association of *PS* and *HS* in university programmes is a 'marriage of convenience'.[6] According to Giere, *PS* is a normative discipline which sets standards of explanation, confirmation, and theory replacement. As such, *PS* is an analysis of methodological problems within present-day science. Giere maintained that such an analysis may be undertaken independently of any specifically historical analysis.

J. J. C. Smart also emphasized the independence of *PS*. He noted that the philosopher of science typically selects examples from *HS* because it is convenient to do so. These examples are selected to illustrate his methodological principles. Smart maintained that fictitious examples would do equally well if the philosopher were sufficiently clever to invent them.[7] Knowledge of *HS* is not required in the practice of *PS*.

Paul Feyerabend agreed that the association of *PS* and *HS* is a marriage of convenience. But he did so for quite different reasons. Feyerabend called into question the entire enterprise of *PS*. He maintained that the analyses of explanation, confirmation, and theory comparison given by the philosopher are of value to neither the scientist nor the historian of science. According to Feyerabend, *PS* is an irrelevant academic exercise which may safely be ignored.[8]

---

[3] Ibid., 5.   [4] Ibid., 10.

[5] The comments of these authors were not given as explicit responses to Kuhn's question.

[6] Ronald Giere, 'History and Philosophy of Science: Intimate Relationship or Marriage of Convenience?' *Brit. J. Phil. Sci.* 24 (1973), 282–97.

[7] J. J. C. Smart, 'Science, History and Methodology', *Brit. J. Phil. Sci.* 23 (1972), 268.

[8] Paul K. Feyerabend, 'Philosophy of Science: A Subject with a Great Past' in *Historical and Philosophical Perspectives of Science*, Minnesota Studies in the

II   PHILOSOPHY OF SCIENCE, HISTORY OF SCIENCE, AND BOHR'S PRINCIPLE OF COMPLEMENTARITY

Kuhn did not accept Feyerabend's recommendation. He insisted that *PS* is an important discipline and that there should be fruitful interaction between *PS* and *HS*. One way in which this might take place would be if *PS* and *HS*, although mutually exclusive on one level, were also complementary disciplines on a higher level. This would be the case, for instance, if *PS* and *HS* conform to the requirements of Niels Bohr's Principle of Complementarity. The Principle of Complementarity is an important feature of the Copenhagen Interpretation of quantum mechanics. On the Copenhagen Interpretation, a hierarchy of language levels is involved in the interpretation of quantum phenomena:

Level 1: Reports of the results of experiments, stated in the language of classical physics.

Level 2: Statements of the values of state-functions for quantum-mechanical systems, which state-functions are interpreted according to Max Born's semantical rule that correlates $|\Psi|^2 dV$ with the probability that a quantum-mechanical system is in volume element dV.

Level 3: 'Descriptions' of the behaviour of quantum-mechanical systems during the intervals between the observations that are made on the system, which 'descriptions' are in terms of the classical concepts of waves and particles.

Although it is possible to describe and predict the results of experiments on quantum-mechanical systems by remaining at Level 2 of the hierarchy, supporters of the Copenhagen Interpretation insist that knowledge of quantum-mechanical systems is incomplete without the superposition of complementary wave and particle pictures. They utilize the Principle of Complementarity as a criterion of completeness for quantum-mechanical interpretations. Born sought to justify acceptance of this criterion by appeal to the principle that 'scientific results should be interpreted in terms intelligible to every thinking man'.[9]

Level 3 decriptions in terms of waves and particles are held to be both mutually exclusive and complementary. Bohr noted that

Philosophy of Science 5, ed. R. Stuewer (Minneapolis: University of Minnesota Press, 1970), 172–83.

[9] Max Born, *Physics in My Generation* (London: Pergamon, 1956), 48.

complementary descriptions are interpretations of experiments conducted to determine the values of different conjugate properties of a quantum-mechanical system. These conjugate properties are related, via the uncertainty relations, to the quantum of action *h*. Louis de Broglie characterized the situation as follows:

> if we wish to establish a correspondence between these two pictures—for example—connect the energy and the momentum of a particle with the frequency and the wavelength which we associate with it, we are led to formulae in which Planck's constant *h* figures in an essential way; this shows that the duality of waves and corpuscles, the necessity of employing two pictures—apparently contradictory—to describe the same phenomena, is closely associated with the existence of the quantum of action.[10]

Bohr recommended application of the Principle of Complementarity in areas far removed from quantum theory. He suggested, for instance, that descriptions of biological phenomena in teleological categories and physicalistic categories, and descriptions of psychological phenomena in volitional categories and behaviourist categories also may be complementary. Indeed, Bohr recommended as a general principle directive of scientific enquiry that mutually exclusive, but complementary, second-order interpretations be sought for first-order subject matter.[11]

There are, however, two important respects in which application of the Principle of Complementarity to history of science and philosophy of science would be inappropriate. In the first place, although experimental arrangements and results can be described in a language neutral with respect to wave–particle dualism, no theory-neutral version of developments in science is available. Every attempt to state what happened in science is already an interpretation and hence either a history of science or a rational reconstruction on behalf of some philosophy of science. In the second place, the choice of which complementary description to apply in a given experimental context is not arbitrary. Given a specific experimental arrangement, either the wave interpretation is appropriate and the particle interpretation is inappropriate, or vice versa.[12] But no such constraints apply to the interpretation of developments in science.

---

[10] Louis de Broglie, *Physics and Microphysics* (New York: Harper, 1960), 125.

[11] Niels Bohr, *Atomic Physics and Human Knowledge* (New York: John Wiley, 1958) 10–12, 76–9, 88–93.

[12] In the case of the diffraction of electrons or photons at a single slit, it is

In view of these important disanalogies, the Principle of Complementarity is not an appropriate directive principle to relate *PS* and *HS*. Given Kuhn's suggestion that *PS* and *HS* may be mutually exclusive *gestalt*-perspectives, a second possible directive principle is that *HS* be practised to the exclusion of *PS*.

Kuhn did not recommend that historical reconstruction be practiced to the exclusion of *PS*. Nevertheless, certain of his remarks support the thesis that *HS* is an appropriate interpretation of developments in science, and that *PS* is not. Kuhn emphasized that many philosophical interpretations of scientific progress fail to capture what is essential in scientific practice. He noted that this failure is not a matter of omitting relevant details. Selection is required in both *PS* and *HS*. But 'the philosopher of science often seems to have mistaken a few selected elements for the whole and then forced them to serve functions for which they may be unsuited in principle and which they surely do not perform in practice'.[13]

These remarks do not preclude the possibility that one day a

---

possible to set up an apparatus to measure accurately either the 'path' of a particle or the diffraction pattern. If the diaphragm is rigidly connected to a massive frame, the transverse component of the position of the particle passing through the diaphragm can be determined by means of the position of the slit. The transverse component of the momentum of the particle may then be determined with only very low accuracy, although the *distribution* of particles striking the photographic plate can be calculated from the Schrödinger equation upon substitution of a specific distance between diaphragm and plate. Bohr pointed out that, if the diaphragm is connected to its frame by means of an elastic spring, 'it should, in principle, be possible to control the momentum transfer to the diaphragm and, thus, to make more detailed predictions as to the direction of the electron path from the hole to the recording point' (Bohr, 'Discussion with Einstein on Epistemological Problems in Atomic Physics' in *Albert Einstein: Philosopher–Scientist*, ed. P. A. Schilpp (New York: Tudor, 1951), 216). This increase in the accuracy with which the path of a particle may be determined is achieved only at the expense of a blurring of the diffraction pattern. Consequently, the experimental arrangement in which an interpretation in terms of 'wave language' is applied (the diaphragm is rigidly connected to the frame) does not readily lend itself to an interpretation in terms of 'particle language', and the experimental arrangement in which an intepretation in terms of 'particle language' is applied (the diaphragm is connected to the frame by means of an elastic spring) does not readily lend itself to an interpretation in terms of 'wave language'. The Principle of Complementarity stipulates a restriction on Level 3 interpretations; the results of a particular experiment may not be interpreted in terms of both a 'particle picture' and a 'wave picture'.

[13] Kuhn, 'The Relations Between the History and the Philosophy of Science'. 14.

non-distorting *PS* will be formulated. However, these remarks do not afford a basis for optimism either.

To claim that *PS* fails to capture what is essential in scientific practice is to appeal to an understanding of what is essential. Presumably this understanding is derived from acquaintance with some *HS*.

Kuhn insisted that *HS* is a discipline in which explanation is achieved. To maintain that philosophical interpretations often distort scientific reality and that historical interpretations sometimes succeed as explanations of this reality is to provide support for the conclusion that *HS* is the more adequate *Gestalt* perspective.

According to Kuhn, a historical interpretation achieves explanation in so far as it displays 'not only facts but also connections between them'.[14] One way in which a historical interpretation might display connections between facts is by deductive subsumption under laws. Kuhn rejected this alternative. He maintained that the Covering-Law Model fails to capture what is essential to historical explanation.

It remains, then, to specify what *is* essential to historical explanation. Kuhn conceded that there is as yet no satisfactory account of historical explanation. As a contribution toward such an account, Kuhn proposed an analogy between historical reconstruction and a special type of puzzle-solving.

III KUHN'S 'PUZZLE-SOLVING' MODEL OF HISTORICAL EXPLANATION

Given a large number of identically-sized square jigsaw pieces and a box within which they are to be arranged, the puzzle-solver seeks to select and juxtapose a subset of these pieces so as to create a recognizable scene. In similar fashion the historian of science seeks to construct from his materials a suitable narrative. A suitable narrative, like a suitable puzzle solution, displays a familiar, if previously unseen, pattern.

According to Kuhn, both puzzle-solving and historical reconstruction are subject to rules. The puzzle-solver must not leave empty spaces in the middle of the box, and the historian must not leave gaps in his narrative. Both puzzle-solver and historian must

---

[14] Ibid., 15.

avoid incongruities—for example the legs of a man must not be conjoined to the body of a sheep, and a cruel tyrant must not spontaneously become a benevolent ruler.[15] Kuhn noted, in addition, that there are rules that are applicable to historical reconstruction but not puzzle-solving. For instance, the historian must select facts for his narrative such that his interpretation does not clash with those facts that are omitted from the narrative. The puzzle-solver, on the other hand, may disregard the unused pieces, provided that he has found a perspicuous arrangement of the remaining pieces.

The most important aspect of the analogy which Kuhn developed is that both successful puzzle solution and successful historical narration achieve a new perspective. Both successful puzzle solutions and successful historical reconstructions are arrangements of a 'given' (puzzle pieces, historical facts) to form 'a familiar, if previously unseen, product'.[16] It is Kuhn's claim that successful historical narratives achieve pattern-recognition in the *Gestalt* sense of 'seeing as'.

Kuhn's Puzzle-Solving Model entails several interesting consequences about the nature of historical reconstruction:

(1) there exist 'historical facts' (a set of jigsaw pieces from among which a puzzle solution is to be achieved);

(2) historical interpretation requires a selection of facts (more pieces are available than can be arranged within the box);

(3) 'similarity-recognition' is a necessary condition of successful historical reconstruction (a proper juxtaposition of pieces demands recognition of similarities of colour, line, and subject matter); and

(4) there are 'solutions' to puzzles about developments in science.

If these consequences are implausible, then the Jigsaw-Puzzle Model is inappropriate. The first consequence is that there exist 'historical facts' which the historian selects and arranges to create a narrative. Among candidates for status as 'historical facts' are records of various kinds. But neither journal articles, nor notebook entries, nor letters, nor photographs of laboratory equipment are like identically-sized puzzle pieces. Documents, for instance, do not come labelled 'authentic' or 'unauthentic'.

---

[15] Ibid., 17.   [16] Ibid., 17.

Authenticity must be established by the historian. And even after questions of authenticity have been settled, the creator of history of science is not in the situation of the puzzle-solver who confronts a set of puzzle pieces that are 'already there'. One cannot create history of science merely by compiling and arranging documents or artefacts. Rather, the historian of science must establish what are the facts from the documents and artefacts that are available to him.

In this reconstruction, perhaps it is descriptions of events, and not documents, which are the analogues of puzzle pieces. Documents do describe events, and the historian can often establish beyond reasonable doubt that events took place in a certain order. However, given a set of events—e_1, e_2, e_3 (under some description)—a sequential ordering of these events is insufficient to create a history of science.

This is because the subject matter of historical narration is descriptions of human *actions*. To describe a human action is not simply to describe an event occurring at a specific time and place. It is necessary to assess the significance of the event by placing it within a context of prior events and subsequent events. Only then is the description a description of an action. Kuhn himself has emphasized this. However, if it is descriptions of human actions that are 'historical facts' and, by analogy, 'puzzle pieces', then it is a strange sort of 'juxtaposition' that is required to create a historical narrative.

Whereas the juxtaposition of puzzle pieces to depict a pattern requires attention to line, colour, and (perhaps also) representational content, the 'juxtaposition' of descriptions of human actions to create a coherent narrative requires attention to temporal relations among the actions (pieces). Human actions are often stages of projects. Consequently, historical reconstruction often involves an ascription of projects to human agents.

Whether or not a historian can assess correctly the significance of an action depends on the position of the action in a sequence of actions. Suppose that Smith executes a project—for instance, the synthesis of compound X—by performing actions a_1, a_2, a_3, a_4, and a_5. At time t_5 the historian has previously recorded events e_1, e_2, e_3, and e_4. He is in a good position to ascertain the significance of e_5—for instance, a melting-point determination of the synthesized compound. At the earlier time t_3, the historian has less information about the project and may not be able to place correctly e_3 as a stage of the projected synthesis of X. At the still earlier time t_1 the historian's knowledge of the project is restricted to his knowledge of e_1—for instance, the construction of a distillation apparatus— perhaps augmented by what he has observed about Smith's behaviour in other contexts. The historian may be quite unable to determine the significance of event e_1. To understand the significance of an event one often needs to have knowledge of subsequent events.

It is because the historian believes an agent to be acting on a particular project that he interprets events e_1, e_2, e_3 . . . to be actions a_1, a_2, a_3 . . . The historian may not be able to specify in advance the next stage of the project. At a given stage of the project a large number of actions may qualify as furthering the project.

Moreover, it is often unclear whether a given action is consistent with the execution of a particular project. There is a penumbra of vagueness associated with the ascription of project verbs. Because of this vagueness, historians may disagree about whether or not an agent really is acting on a given project. For instance, historian 1 may note that Jones did a_1 at t_1, a_2 at t_2, and a_3 at t_3, and conclude that Jones is acting on project Q. Historian 2 may object that the sequence of actions a_1, a_2, a_3 does not point toward project Q. To which historian 1 may reply that certainly the sequence b_1, b_2, b_3 would be actions within project Q and that a_1 and a_2 resemble b_1 and b_2 in certain respects. To which historian 2 may counter that b_3 occurs only when certain types of circumstances are present, and that no such circumstances were present when a_3 occurred.

Such debates depend upon judgements about resemblance and difference. Suppose the point at issue is whether a statesman is pursuing a policy of reconciliation with his country's recent enemy. Extreme cases are easily decided. If it can be shown that the statesman arranged for the assassination of twelve high officials of the other country, then he most certainly is not pursuing a policy of reconciliation. But the range of actions that are consistent with a project of reconciliation is wide indeed, and reasonable men may disagree about the status of such actions.

Arthur Danto has emphasized that project-ascriptions have a peculiar logic.[17] They cannot be falsified in a simple, direct manner. It may be correct to state that 'Smith is working on the synthesis of compound X' even at those times when he is puzzling instead about the structure of compound Y. Clearly Smith must perform some actions to qualify as working on the synthesis of X, but these actions need not be continuous, and it is difficult to determine if, and at what point, he has abandoned the project. Certainly, it may be true that 'Smith is working on the synthesis of X' even though X is never produced. We are in the odd situation of having to admit that the meaning of a project-ascription depends on a description of a future state of affairs, but that the ascription is not falsified simply because that future state is not realized.

It might be objected on behalf of the Puzzle-Solving Model that an action 'points toward' other actions (antecedent and subsequent) in the same way that a puzzle piece 'points toward' its proper neighbours in virtue of its line, colour, or representational content. From this perspective, the temporal relatedness of historical facts is just one additional type of relatedness to which the operation of similarity-recognition must be applied.

Similarity-recognition is an important aspect of the Puzzle-Solving Model. It is a necessary condition of both puzzle solution and historical explanation. Kuhn declared that

if history is explanatory, that is not because its narratives are covered by general laws. Rather it is because the reader who says 'Now I know what happened' is simultaneously saying 'Now it makes sense; now I understand; what was for me previously a mere list of facts has fallen into a recognizable pattern.'[18]

No doubt historical reconstruction does involve similarity-recognition.[19] And Kuhn may be correct that similarity-recognition is unanalysable in the sense that necessary and sufficient conditions

---

[17] Arthur C. Danto, *Analytical Philosophy of History* (Cambridge: Cambridge University Press, 1965), 143–81, 233–56.

[18] Kuhn, 'The Relations Between the History and the Philosophy of Science', 17–18.

[19] Kuhn noted that similarity-recognition is also basic to the practice of science. 'Normal science' proceeds by application of paradigm-exemplars. In these applications 'scientists model one problem solution on another without at all knowing what characteristics of the original must be preserved to legitimate the process' (ibid., 17).

cannot be specified for the similarity of two situations.[20] However, there are important differences in the ways in which puzzle-solver and historian utilize their capacities to recognize similarities. The puzzle-solver who has recognized a pattern and has arranged his pieces to implement this recognition has solved the puzzle. Of course, there are other selections and arrangements that can be made, and it is possible that some of these arrangements may also evoke similarity-recognition. But there is no particular reason for the puzzle-solver to seek alternative solutions.

The situation of the historian is quite different. At a given time the historian may achieve similarity-recognition by superimposing a project-ascription upon a set of human actions. However, this achievement lacks the finality of a puzzle solution. At a subsequent time the historian may correctly conclude that his initial project-ascription was wrong. If the actions under examination are stages of an ongoing project, subsequent events may indicate that the actions initially considered are stages of a different project (e.g. the project was not to synthesize compound X but rather to assess the toxicity of compound Y). But even if the actions under examination are stages of a completed (or abandoned) project, additional historical study may disclose that events initially ignored were important, or that events initially considered had been wrongly interpreted.

It is indisputable that historians disagree about the reconstruction of the past. In many cases, the divergent interpretations are different ways of reading an agreed-upon sequence of events. What is at stake is assessment of the significance of these events, and this assessment seems to depend on the predispositions, sensitivities, and beliefs of historians. (Consider, for example, the interpretations of the work of William Gilbert by Zilsel[21] and Hesse,[22] the interpretations of the work of William Harvey by Whewell,[23] Nordenskiöld,[24] and Pagel,[25] the interpretations of the

[20] Ibid., 17.

[21] Edgar Zilsel, 'The Origins of William Gilbert's Scientific Method', *JHI* 2 (1941), 1–32.

[22] Mary B. Hesse, 'Gilbert and the Historians', *Brit. J. Phil. Sci.* 11 (1960), 1–10, 130–42.

[23] William Whewell, *History of the Inductive Sciences* (3rd edn., 1857; London: Cass, 1967), 3. 332–4.

[24] Eric Nordenskiöld, *The History of Biology* (New York: Tudor, 1928), 114–18.

[25] Walter Pagel, *William Harvey's Biological Ideas* (Basel: S. Karger, 1967).

work of Francis Bacon by Herschel,[26] Dijksterhuis,[27] Butterfield,[28] Farrington,[29] and Rossi,[30] and the interpretations of the work of Isaac Newton by Hessen,[31] Clark,[32] Manuel,[33] and Westfall.[34]

That there has been diversity in the interpretation of developments in the history of science is a factual claim, not dependent on the truth of any of the various forms of historical relativism. However, there is nothing in the Puzzle-Solving Model to lead one to expect this sort of diversity.

Of course, the Puzzle-Solving Model may be salvaged by altering the nature of puzzle-solving so that it more closely matches the activity of the historian. Perhaps the puzzle pieces are of such a nature that a number of plausible arrangements are possible, each likely to be uncovered only by a puzzle-solver with appropriate experience. But such alteration of the model would be *ad hoc*. The intent of the analogy to puzzle-solving was to clarify the nature of historical reconstruction. It would defeat this intent to appeal to an antecedent understanding of historical reconstruction in order to correct deficiencies in the analogy.

---

[26] John F. W. Herschel, *A Preliminary Discourse on the Study of Natural Philosophy* (1830; New York: Johnson, 1966), 104–17.

[27] E. J. Dijksterhuis, *The Mechanization of the World Picture* (Oxford: Clarendon Press, 1961), 396–403.

[28] Herbert Butterfield, *The Origins of Modern Science* (New York: Macmillan, 1957), 96–116.

[29] Benjamin Farrington, *Francis Bacon: Philosopher of Industrial Science* (London: Lawrence and Wishart, 1951).

[30] Paolo Rossi, *Francis Bacon: From Magic to Science* (London: Routledge & Kegan Paul, 1968).

[31] B. Hessen, *The Social and Economic Roots of Newton's 'Principia'* (New York: Howard Fertig, 1971).

[32] G. Clark, *Science and Social Welfare in the Age of Newton* (Oxford: Clarendon Press, 1970).

[33] F. Manuel, *A Portrait of Isaac Newton* (Cambridge, Mass.: Harvard University Press, 1968).

[34] R. S. Westfall, *Never at Rest* (Cambridge: Cambridge University Press, 1980).

3

More Than a Marriage of Convenience

IT has not proved helpful to superimpose a *Gestalt*-perspective analogy upon *PS* and *HS*. The analogy is inappropriate because *PS* and *HS* are not mutually exclusive interpretations of developments in science. Indeed, there are a number of reasons to believe that the association of *PS* and *HS* is a genuine marriage in which each partner is affected by changes in the other.

It seems clear that to practice *HS* it is necessary to judge the significance of developments within science. The historian of science, *qua* historian, renders judgements of importance about the evidence available to him. These judgements reflect his understanding of what counts as science and what kinds of factors affect its development. This is not to say that commitment to a fully articulated *PS* is a necessary condition of historical reconstruction. Nevertheless, the selection, organization, and interpretation of historical data presuppose principles whose origin is a *PS*.

I CONFIRMATION AND HISTORICAL ENQUIRY

It seems clear as well that historical enquiry is required in the practice of *PS*. If *PS* is taken to be a descriptive discipline whose aim is to uncover the evaluative standards that have been effective in science, then without question historical enquiry is required in the practice of *PS*. But historical enquiry is also required in the practice of a prescriptive *PS*.

This is true even within a logicist version of prescriptive *PS*. In its logicist form, *PS* is a discipline in which scientific laws and theories are reformulated in the patterns of formal logic and questions of confirmation or explanation are dealt with as problems in applied logic. On the Logical Reconstructionist view,[1] dominant within *PS* from 1940 until perhaps 1970, to formulate an evaluative criterion for *PS* is to specify an 'explication' of a

---

[1] Proponents of the Logical Reconstructionist version of *PS* include Rudolf Carnap, Carl Hempel, R. B. Braithwaite, and Ernest Nagel.

corresponding epistemological term. In the case of the qualitative relation between a hypothesis and a proposition reporting evidence, the corresponding epistemological term is 'qualitative confirmation'. Given hypothesis *h* and evidence *e*, the required explication is a definition of '*e* confirms *h*' in terms of logical concepts such as consistency and entailment.

If a proposed explication is satisfactory, then it captures a pre-analytic usage within scientific practice. This pre-analytic usage is recorded in a *HS*. But of course an explicated concept is not a summary of actual usage. Rather, a definition of qualitative confirmation is typically put forward as a refinement of actual usage. Failure of certain cases of evaluative practice in science to conform to the requirements of a proposed concept of confirmation need not discredit the definition. But a substantial segment of scientific evaluative practice must conform, approximately at least, to the explicated concept. Otherwise it is not 'confirmation' in its scientific setting that has been explicated. It would be a rash Logicist indeed who systematically dismissed as irrelevant a marked divergence between application of his criterion and the course of scientific practice. It seems clear that justificatory arguments about proposed evaluative criteria for a prescriptive *PS* *ought* to involve reference to *HS*.

Hempel's 'Satisfaction Criterion' of qualitative confirmation

Historical considerations have proved relevant to the philosopher's search for an appropriate criterion of qualitative confirmation in a further significant respect. The original Logical Reconstructionist programme was to explicate 'qualitative confirmation' in terms of the syntactic properties of a specified formal language.

Carl Hempel suggested in 1945 that a proper definition of '*e* confirms *h*' should fulfil the following four requirements:[2]

(1) *Entailment Condition*—If *O* entails *H*, then *O* confirms *H*.

(2) *Special Consequence Condition*—If *O* confirms *H*, then *O* confirms every logical consequence of *H*.

(3) *Equivalence Condition*—If *O* confirms *H*, then *O* confirms every *H'* logically equivalent to *H*.

---

[2] Carl Hempel, 'Studies in the Logic of Confirmation' in *Aspects of Scientific Explanation* (New York: Free Press, 1965), 30–5. This article is a revised version of an article of the same title that appeared in *Mind 54* (1945), 1–26; 97–121.

(4) *Consistency Condition*—Every logically consistent O is logically compatible with the class of all Hs that it confirms.

Hempel noted that it would not do to add to these requirements a 'Converse Consequence Condition'.

(5) *Converse Consequence Condition*—If O confirms a logical consequence of H, then O confirms H.

If this Converse Consequence Condition were adopted as a requirement along with (1), (2), and (3) above, then observation report O would confirm any H whatsoever.[3] Hempel also observed that the Consistency Condition, as stated, implies the following two requirements for non-self-contradictory observation reports:

(4*a*) If O and H are not logically compatible, then O does not confirm H.

(4*b*) If H_1 and H_2 are logically incompatible, then O confirms neither.

Hempel acknowledged that there are objections to requirement (4*b*). In particular, a given set of values $x_i y_i$ may be a substitution instance of each of two logically incompatible mathematical functions. We would want to say that $(x = 2, y = 3)$ confirms H_1—$(y = 2x - 1)$—despite the fact that H_1 and H_2—$(y = x^2 - 1)$—are not logically compatible. Hempel stated that it would be of value to work out a theory of confirmation in which the weaker requirement (4*a*) replaces requirement (4). However, the definition of confirmation which he himself developed is based on the stronger requirement (4).

Hempel's 'Satisfaction Criterion' of confirmation is formulated for languages whose structure is precisely specified. Such languages have the following ingredients:

terms designating more or less directly observable attributes of things and events—property terms and relational terms
constants designating individuals—a, b, c . . .
variables—x, y . . .
sentence connectives— · , v, ∼ . . .
Quantifiers—(x) . . . , $(\exists x)$ . . .
rules of sentence formation

---

[3] Since, for any O and any H
    (1) O confirms O, by the Entailment Condition,
    (2) O confirms $(O \cdot H)$, by the Converse Consequence Condition, and
    (3) O confirms H, by the Special Consequence Condition.

rules of deductive inference

observation sentences—assert that a given object has a particular property, or that given objects stand in a particular relationship —e.g. *Pa, Rab* . . .

hypotheses—any sentence, quantified or particular, which can be formed from the above ingredients according to the rules of sentence formation.

The Satisfaction Criterion is expressed in terms of the concept of the 'development' of a hypothesis. Roughly speaking, the development of a hypothesis H for a finite class of objects C states what H would assert if there existed only the objects of C.[4]

Given the notion of development, and a language with the structure outlined above, Hempel then stipulated the following definition of 'O directly confirms H':

> O directly confirms H if O entails the development of H for the class of objects mentioned in O.

For example, consider observation report O_1 about the blackness and 'ravenhood' of individuals a and b—$O_1 = $ '$(Ra \cdot Ba) \cdot (\sim Rb \cdot Bb)$'. O_1 'directly confirms' hypothesis H_1 that 'All ravens are black'—'(x) $(Rx \supset Bx)$'—because O_1 entails the development of the hypothesis for the class a, b, namely '$(Ra \supset Ba) \cdot (Rb \supset Bb)$'.[5]

Hempel next defined 'O confirms H' in terms of 'O directly confirms H':

> O confirms H if O directly confirms each member of a class of sentences C, which C entails H.

For example, $H_2 = (Rc \supset Bc)$ is confirmed by O_1, because O_1 directly confirms H_1 and H_1 entails H_2. Hempel declared that

the criterion expressed in these definitions might be called the *satisfaction criterion of confirmation* because its basic idea consists in construing a hypothesis as confirmed by a given observation report if the hypothesis is satisfied in the finite class of those individuals which are mentioned in the report.[6]

---

[4] Hempel, 'Studies in the Logic of Confirmation', 36.

[5] O_1 entails $(Ba \cdot \sim Rb)$ which, in turn, entails $(Ba \vee \sim Ra) \cdot (\sim Rb \vee Bb)$ by the Addition Rule of the Sentential Calculus—

$$\frac{p}{\therefore p \vee q}$$

And $(Ba \vee \sim Ra) \cdot (\sim Rb \vee Bb)$ is logically equivalent to $(Ra \supset Ba) \cdot (Rb \supset Bb)$.

[6] Hempel, 'Studies in the Logic of Confirmation', 37.

Rudolf Carnap, who shared Hempel's commitment to the Logical Reconstructionist view of *PS*, showed in 1950 that, of Hempel's four conditions for a proper definition of '*e* confirms *h*', only the Equivalence Condition can be accepted without qualification. Carnap demonstrated that the Entailment Condition must be restated to exclude certain cases that may arise in infinite language systems.[7] He also produced a countercase to show that the Special Consequence Condition is not valid.[8] But his major complaint was that Hempel did not take seriously his own acknowledgement of objections to the Consistency Condition. The Consistency Condition entails that

(4*b*) If H_1 and H_2 are logically incompatible, then O confirms neither.

Hempel had conceded that a given set of measured values may confirm two or more logically incompatible quantitative hypotheses. Despite this, Hempel developed the Satisfaction Criterion of confirmation on the assumption that (4*b*) holds.

Carnap suggested that Hempel may have proceeded in this manner because the results of physical measurements cannot be formulated in the simple artificial language for which the Satisfaction Criterion was developed. However, Carnap found a countercase to the Consistency Condition which can be formulated within Hempel's language. Consider a finite population from which sample *s* has been withdrawn. Let h_1 and h_2 state different estimates of the relative frequency of individuals in the sample which have property *P*. Hypotheses h_1 and h_2 are logically incompatible.

If h_1 and h_2 are both very close to the actual statistical distribution—*i*—of individuals which have *P* within the total finite population, then *i* confirms both h_1 and h_2. For example, if h_1 states that 80 individuals in a sample of 100 have *P*, and h_2 states that 79 individuals in the sample have *P*, evidence that $i = 7,950$ individuals with *P* in a population of 10,000 would confirm both h_1 and h_2. Thus requirement (4*b*) cannot be retained.[9]

'Historical' views of confirmation

There is a further complication. It has been assumed to this point

---

[7] Rudolf Carnap, *Logical Foundations of Probability* (Chicago: University of Chicago Press, 1950), 473–4.

[8] Ibid., 474–5; 393–5.   [9] Ibid., 476–7.

that criteria of evidential support state a relation between a hypothesis and its instances. On this assumption, it matters not whether the hypothesis or the evidence is presented first. As Hempel put it, 'from a logical point of view, the strength of the support that a hypothesis receives from a given body of data should depend only on what the hypothesis asserts and what the data are'.[10]

A number of philosophers of science have argued, on the contrary, that the temporal relationship between hypothesis and evidence is of great importance. Karl Popper, for instance, maintained that it is not instances, *per se*, which provide evidential support for hypotheses, but *tests*. A test, according to Popper, is a serious attempt to falsify a hypothesis.[11]

Whether or not instance '$(Aa \cdot Ba)$' counts as a test of hypothesis '(x) $(Ax \supset Bx)$' may depend on our background knowledge. Suppose a new plant hybrid is created. After plants Pa, Pb, Pc, and Pd have been produced, it is observed that each of the four has white flowers. A scientist may put forward the hypothesis that all such plants have white flowers. Since '$(Pa \cdot Wa)$' is part of the background knowledge available at the time the hypothesis is formulated, it does not qualify as a test of the hypothesis. Only plants other than the first four are bona fide tests of the hypothesis.

This understanding of evidential support does accord with the intuitions of many scientists. Scientists do apply hypotheses to instances not initially considered. And they are impressed by successful prediction. Even if the hypothesis in question seems to be an *ad hoc* numerical correlation, application of the correlation to new cases may convert sceptics into believers.

Consider the Titius–Bode correlation in which observed planetary distances are correlated with 'suitably adjusted' terms of the geometrical series 3, 6, 12, 24 . . . (see Table 1). The correlation achieved for the first six planets is quite close. But since these distances were known before the correlation was suggested, and since there was no planet known between Mars and Jupiter, most

[10] Hempel, *Philosophy of Natural Science* (Englewood Cliffs: Prentice–Hall, 1966), 38.

[11] Karl Popper, *Conjectures and Refutations* (New York: Basic Books, 1962), 36–7, 241–2; 'The Aim of Science', in *Objective Knowledge* (Oxford: Oxford University Press, 1972), 192–3.

Table 1. Titius–Bode Correlation

| | 4 | 4 | 4 | 4 | 4 | 4 | 4 | 4 | 4 |
	0	3	6	12	24	48	96	192	384
Calculated	4	7	10	16	28	52	100	196	388
Planet	Merc.	Venus	Earth	Mars	?	Jup.	Sat.	(Ur.)	(Nep.)
Observed	3.9	7.2	10.0	15.2		52.0	95.4	192.0	301.0

scientists dismissed the correlation as an 'after the fact' numerical coincidence. But in 1781, William Herschel discovered what turned out to be a new planet beyond Saturn. Its distance from the sun was determined to be 192, in good agreement with the next term in the series. A search was begun for the 'missing planet' at distance 28. In 1801 the first asteroid was discovered. The Titius–Bode correlation had achieved 'predictive success' again. Next, both Adams and Leverrier used the distance value 388 in their calculations of the position of a trans-Uranic planet. Neptune was discovered in the region predicted. However, its distance from the sun turned out to be 301, a value not in good agreement with the value given by the Titius–Bode series.[12] The history of the changing fortunes of the Titius–Bode relation reveals that scientists sometimes place great emphasis on 'new' cases.

In an essay published in 1974, Alan Musgrave contrasted the Logical Reconstructionist view of qualitative confirmation with three 'historical' theories of confirmation.[13] The three theories are historical because they make estimation of evidential support depend on historical enquiry.

The first of these theories is a 'strictly temporal' view of evidential support. On this view, facts 'known to science' prior to the formulation of hypothesis *H* do not provide evidential support for *H*. The criterion for evidential support is that

> *e* supports *H* provided that
> (1) *H*, in conjunction with statements of initial conditions and boundary conditions, implies *e*, and

[12] Adams and Leverrier accurately located Neptune's position against the background stars in spite of the fact that they based their calculations on an erroneous value for the planet's distance from the sun. See, for instance, Morton Grosser, *The Discovery of Neptune* (Cambridge: Harvard University Press, 1962), Chs. 5–7.

[13] Alan Musgrave, 'Logical versus Historical Theories of Confirmation', *Brit. J. Phil. Sci.* 25 (1974), 1–23.

(2) e is not known to science prior to the formulation of H.

The concept 'known to science' is vague, and historians may disagree on the time at which an experimental result became 'known to science'. But there is a further, more important, difficulty. Musgrave indicated the difficulty by contrasting three possible historical sequences.[14] In sequence 1, (see Figure 1) $e_2 \ldots e_n$ support H, because H 'implies' $e_2 \ldots e_n,$[15] and $e_2 \ldots e_n$ are

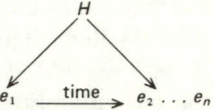

Fig. 1. Musgrave's Sequence 1

formulated subsequently to H. In sequence 2 (see Figure 2), $e_2 \ldots e_n$ do not support H, because $e_2 \ldots e_n$ are presumed to be 'known

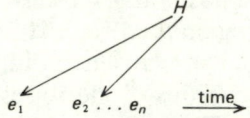

Fig. 2. Musgrave's Sequence 2

to science' at the same time H is formulated. In sequence 3 (see Figure 3), H_2, in conjunction with statements of initial conditions,

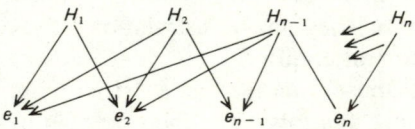

Fig. 3. Musgrave's Sequence 3

implies e_2, whereas H_1, in conjunction with the same statements, implies that e_2 is not the case. On the strictly temporal view of evidential support, $e_1 \ldots e_n$ do not support H_n, despite the fact that H_n 'explains' (deductively subsumes) the predictive failure of $H_1 \ldots H_{n-1}$.

Musgrave maintained that this is a counterintuitive result. By

---

[14] Ibid., 9–11.
[15] That is, H, in conjunction with statements of initial conditions and boundary conditions, implies $e_2 \ldots e_n$.

way of example, he called attention to the relation between the Michelson–Morley experiment and the Theory of Special Relativity. Musgrave insisted that the M–M result supports Special Relativity Theory even though the M–M result was 'known to science' when Einstein formulated the theory.

A second historical theory of confirmation was suggested by Elie Zahar.[16] Musgrave labelled it a 'heuristic' view of evidential support. Zahar suggested that evidential support depends not on what was 'known to science' but on those facts wh ch were taken into account in the formulation of the hypothesis.[17] Thus because Einstein did not take into account the M–M result in formulating the Theory of Special Relativity, the M–M result *does* support the theory.

Musgrave objected to the subjectivist implications of Zahar's heuristic view. On Zahar's view, evidential support may depend on biographical details about the life of the theorist. It is possible that a theorist may be aware of a fact without referring to it in the formulation of a theory. And that fact may turn out to be implied by the theory. Presumably, Einstein's formulation of Special Relativity Theory is a case of this kind.[18] But is the theorist's own testimony always the final court of appeal? Is the theorist always, or even usually, a competent judge of his own procedures? Perhaps there is a less subjectivist alternative view of evidential support.

A third historical theory of confirmation was advanced by Imre Lakatos.[19] Lakatos suggested that evidential support for H be determined relative to some 'touchstone' hypothesis H_T which is a serious rival to H. The essential question is whether e supports H more than H_T. For H to receive evidential support relative to H_T, H must have excess empirical content, that is there must be potential falsifiers of H which are not also potential falsifiers of H_T.[20]

---

[16] Elie Zahar, 'Why Did Einstein's Programme Supersede Lorentz's?' *Brit. J. Phil. Sci.* 24 (1973), 95–123, 223–62.

[17] Ibid., 103.

[18] Einstein himself maintained on more than one occasion that the Michelson–Morley result did not play a role in his development of Special Relativity Theory. Gerald Holton has provided a detailed examination of Einstein's various pronouncements about the origin of Special Relativity Theory (Holton, 'Einstein, Michelson, and the "Crucial" Experiment', *ISIS* 60 (1969), 133–97).

[19] Imre Lakatos, 'Changes in the Problem of Inductive Logic' in *Inductive Logic*, ed. I. Lakatos (Amsterdam: North–Holland, 1968).

[20] Ibid., 376–7.

Lakatos's theory of selective confirmation is more objective than those theories which refer evidential support to what is 'known to science' or what was taken into account by the scientist who formulated H. Lakatos's criterion for evidential support is that

  e supports H provided that
  (1) H, in conjunction with statements of initial conditions and boundary conditions, implies e, and either
  (2a) H_T, in conjunction with the same statements, implies that e is not the case (even if e was known prior to H), or
  (2b) H_T, in conjunction with the same statements, implies neither e nor $\sim e$.

Lakatos's theory provides a satisfactory analysis of sequence 3 above. e_n supports H_n because H_n 'implies' e_n whereas H_{n-1}, the touchstone hypothesis, 'implies' that e_n is not the case. Under these conditions e_n supports H_n even though e_n was known before H_n and might have played a role in the formulation of H_n.

Some examples of sequence 3 in the history of science are listed in Table 2. In each case, e supports H on Lakatos's theory of confirmation.

Table 2. Examples of Musgrave's Sequence 3 in the History of Science

e	H	H_T
Weight calx > weight corresponding metal	Oxygen Theory of Combustion	Phlogiston Theory
No pores found in septum of heart (Vesalius)	Harvey's Theory of Circulation	Galen's Theory
$c_{air} > c_{H_2O}$	Maxwell's Electro-dynamic Theory	Newtonian Corpuscular Theory
Michelson–Morley experiment	Special Relativity	Ether Theory
Anomalous Perihelion of Mercury	General Relativity	Newtonian Mechanics

Lakatos's criterion of evidential support is a logicist criterion. It stipulates that satisfaction of a set of logical relations is a sufficient condition of the selective qualitative confirmation of a hypothesis. However, every application of the criterion requires identification of an appropriate touchstone hypothesis, and this identification is achieved by reference to *HS*. If Lakatos is correct that qualitative confirmation is a three-term relation among two hypotheses and a proposition reporting evidence, then there is yet another respect in which *HS* is involved in the practice of *PS*.

II THEORY APPRAISAL AND HISTORICAL ENQUIRY

Historical enquiry is required as well in the application and justification of criteria of theory appraisal. Ernan McMullin has emphasized that a good scientific theory is 'resilient'. It possesses 'a quality of metaphor . . . which suggests to the scientist how its conceptual structures can be further developed to derive new results or to meet new challenges'.[21] If McMullin is correct that theory appraisal ought to include estimation of fertility—proven and/or potential—then there is a further respect in which reference to developments within *HS* is important within a prescriptive *PS*.

Herschel on 'undesigned scope'

McMullin's emphasis on the resilience of theories echoes John Herschel's insistence that a display of 'undesigned scope' provides particularly important evidential support for a theory. Herschel suggested both psychological and logical indicators of undesigned scope. Psychologically considered, instances of undesigned scope are 'unexpected and peculiarly striking confirmations'.[22] Logically considered, instances of undesigned scope extend the range of application of a theory to *additional kinds* of phenomena.[23]

Some examples of undesigned scope cited by Herschel display psychological novelty but no appreciable increase in range of application. He noted, for instance, that one consequence of Fresnel's theory of diffraction is that there should be a bright spot in the centre of the shadow cast by a metal disk illumined by a point source.[24] Despite the fact that this consequence is a violation of common-sense expectations, it was confirmed experimentally. Herschel took the 'unexpectedness' of this result as elevating its value above that of ordinary confirming instances. But this

[21] Ernan McMullin, 'History and Philosophy of Science: A Marriage of Convenience?' in *Boston Studies in the Philosophy of Science* 32, ed. R. S. Cohen *et al.* (Dordrecht: D. Reidel, 1976), 597.

[22] John F. W. Herschel, *A Preliminary Discourse on the Study of Natural Phenomena* (London: Longman, etc., 1830), 171.

[23] Ibid., 172.　　　　　　　　　　　　[24] Ibid., 33–4.

confirmation, although 'striking', could hardly be said to have markedly increased the subsumptive power of Fresnel's theory.

Other examples of undesigned scope cited by Herschel do appear to increase the subsumptive power of a theory. One such case is Laplace's application of the theory of heat to a discrepancy between calculated and observed velocities of sound.[25] The theory of heat implies that heat is produced upon compression of an elastic fluid. Laplace reasoned that this effect occurs during the propagation of sound waves. He calculated the amount of heat produced in this manner, revised the calculated value of the velocity of sound, and achieved agreement with the observed velocity.

Laplace's achievement was unexpected in the sense that the theory of heat was not formulated to deal with problems about the propagation of sound. In addition, it may be argued that Laplace succeeded in extending the theory of heat to a new range of phenomena.

However, Herschel failed to specify the conditions under which a confirmation counts as an instance of a *new type* of phenomena. Presumably, Laplace's recalculation of the velocity of sound qualifies, whereas an application of Snel's law of refraction to a new pair of fluids would not qualify. But surely a given theory has a given set of deductive consequences. It is only scientists' recognition that these consequences are realized empirically that is subject to change. In some cases, recognition of the consequences of a theory counts as uncovery of undesigned scope, and in other cases it does not. It all depends on the psychological impact of the confirmation in question. Whether or not a particular confirming instance is 'unexpected and peculiarly striking' is determined by the reactions of scientists over time, and it is the historian of science who records these reactions.

Reduction and history of science

William Whewell, writing in the mid-nineteenth century, concluded from extensive study of historical developments that progress in science is an incorporation of past results into present theories.

---

[25] Ibid., 171–2.

Whewell superimposed a Tributary–River pattern upon *HS* and concluded that the pattern fits.[26]

Many philosophers of science have agreed with Whewell that scientific progress is growth by incorporation. Ernest Nagel, for instance, has observed that it often happens that after a scientific theory has been introduced it is explained by reference to a more inclusive theory. He declared that 'the phenomenon of a relatively autonomous theory becoming absorbed by, or reduced to, some other more inclusive theory is an undeniable and recurrent feature of the history of modern science'.[27]

Nagel suggested that the incorporation of classical thermo-dynamics into statistical mechanics is a paradigm case of reduction.[28] Maxwell and Boltzmann had deduced laws about the macroscopic behaviour of a gas from premises about the microstructure of the gas.

They proved that the Ideal Gas Law— $\dfrac{PV}{T} = k$ —is a deductive consequence of premises that include the following:

(1) an ideal gas is composed of minute perfectly elastic spheres which collide with one another and the walls of the container;

(2) the collisions of these molecules obey Newton's laws of motion;

(3) the distribution of molecular velocities is given by a statistical assumption about randomness; and

(4) the kinetic energy of the ensemble of molecules is directly proportional to the absolute temperature of the gas— $E = \dfrac{3}{2} k T$.

Reflecting on this achievement, Nagel concluded that reduction is achieved if, and only if, certain conditions are fulfilled. He subdivided these necessary and jointly sufficient conditions into formal conditions and empirical conditions.[29]

---

[26] William Whewell, *History of the Inductive Sciences* 3rd edn. 1857; (London: Cass, 3 vols., 1967).

[27] Ernest Nagel, *The Structure of Science* (New York: Harcourt, Brace & World, 1961), 336–7.

[28] Ibid., 342–66; 'The Meaning of Reduction in the Natural Sciences' in *Readings in Philosophy of Science*, ed. P. Weiner (New York: Charles Scribner, 1953), 535–45.

[29] Nagel, *The Structure of Science*, 345–66.

Formal Conditions for Reduction of T_1 to T_2

1. Connectability: for each term, which occurs in T_1 but not in T_2, there is a connecting statement which links the term with the theoretical terms of T_2.[30] Nagel emphasized that this condition is fulfilled only if the meanings of the respective terms are fixed by appropriate rules of usage.

2. Derivability: the experimental laws of T_1 are deductive consequences of the theoretical assumptions of T_2.

In addition to the formal conditions of connectability and derivability, Nagel proposed two empirical conditions for reduction:

Empirical Conditions of Reduction of T_1 to T_2

3. Empirical support: the theoretical assumptions of T_2 are supported by evidence over and above the evidence that supports T_1.

4. Fertility: the theoretical assumptions of T_2 are suggestive of further development of T_1. Nagel noted that the further development of T_1 often takes the form of an augmentation or correction of the body of laws included within T_1. In the case of thermodynamics, the relationship to statistical mechanics suggested modification of the hypothesis about a microstructure to incorporate intermolecular forces and effects arising from the finite volume of the spheres. In 1873, van der Waals suggested that the Ideal Gas Law be replaced by $(P + \frac{a}{V^2})\ (V - b) = k\ T$, where a is a measure of intermolecular attractive forces and b is a correction term to account for the finite volume of the molecules. The van der Waals equation reproduces the observed pressure-volume-temperature behaviour

---

[30] In the reduction of thermodynamics to statistical mechanics, the connecting statement '$E = \frac{3}{2} k\ T$' is a physical hypothesis subject to confirmation upon independent determination of values of E and T. '$E = \frac{3}{2} k\ T$' states an equivalence of meaning between 'temperature', a term of the secondary science, and 'kinetic energy', a term of the primary science.

Kemeny and Oppenheim suggested that connecting statements between the terms of reduced and reducing branches of science always have biconditional form (John Kemeny and Paul Oppenheim, 'On Reduction', *Phil. Stud.* 7 (1956), reprinted in *Readings in the Philosophy of Science*, ed. B. A. Brody (Englewood Cliffs; Prentice–Hall, 1970), 310). Nagel replied that reduction may also be achieved by the use of connecting statements of conditional form—'If t_1, then t_2'—where t_2 and t_1 are terms of the secondary and primary sciences, respectively (*The Structure of Science*, 355 n.).

of gases over a wider range of values than does the Ideal Gas Law.

Whether or not there has occurred further development of a theory whose reduction is in question can be established only by appeal to *HS*. Thus there is yet another point at which historical enquiry is required in the practice of *PS*.

But even before the question of fertility is raised, historical enquiry is required to establish the specific content of theories T_1 and T_2. In papers published in the 1960s, Paul Feyerabend argued that various prima-facie cases of 'reduction' from *HS* fail to satisfy Nagel's formal conditions for reduction. One such example is the supposed reduction of Galilean physics to Newtonian physics. Feyerabend noted that Nagel's condition of derivability is not fulfilled in this case. A basic law of Galilean physics is that the vertical acceleration of falling bodies is constant over any finite vertical interval near the earth's surface. But this law cannot be deduced from the laws of Newtonian physics. In Newtonian physics the gravitational attractive force, and hence the mutual acceleration of two bodies, increases with decreasing distance. The Galilean law could be derived from Newtonian laws only if the ratio $\dfrac{\text{distance of fall}}{\text{radius of earth}}$ were 0. But in cases of free fall, this ratio never is equal to zero. The Galilean relation does not follow logically from the laws of Newtonian mechanics.[31]

A second example is the supposed 'reduction' of Newtonian mechanics to General Relativity Theory. Feyerabend conceded that under certain limiting conditions, the equations of Relativity Theory yield values that approach those calculated within Newtonian mechanics. But this does not suffice to establish the reduction of Newtonian mechanics to General Relativity Theory. Nagel's condition of definability is not fulfilled in this case. Consider the concept 'length'. In Newtonian mechanics, length is a relation that is independent of signal velocity, gravitational fields, and the motion of the observer. In Relativity Theory, length is a relation whose value *is* dependent on signal velocity, gravitational fields, and the motion of the observer. The transition from Newtonian mechanics to Relativity Theory involves a change of meaning of spatio-temporal concepts. Feyerabend maintained that 'classical

---

[31] Paul Feyerabend, 'Explanation, Reduction, and Empiricism' in *Minnesota Studies in the Philosophy of Science* 3, ed. H. Feigl and G. Maxwell (Minneapolis: University of Minnesota Press, 1962), 46–8.

length' and 'relativistic length' are incommensurable notions,[32] and that Newtonian mechanics is not reducible to General Relativity Theory. He also maintained that classical mechanics cannot be reduced to quantum mechanics,[33] and that classical thermodynamics cannot be reduced to statistical mechanics.[34] In addition, David Hull has advanced convincing arguments to show that fulfilment of the Nagelian formal conditions for reduction is unlikely for any of a number of versions of Mendelian genetics and molecular genetics.[35]

Hilary Putnam suggested that Nagel's Theory of Reduction can be protected gainst Feyerabend's criticism by means of a minor modification. We need only specify that it is a suitable approximation of the old theory that is deducible from the new one. For example, classical geometrical optics is not deducible from (and hence is not reducible to) electromagnetic field theory, but a suitable approximation to classical geometrical optics is so deducible. Putnam emphasized that all that should be expected from a theory of reduction is a set of criteria for the reduction of an approximate version of the replaced theory.[36]

Feyerabend replied that this is not enough. Given G = geometrical optics, and E = electromagnetic field theory, Putnam had conceded that G and E are inconsistent, but had insisted that there exists a theory G' which does follow from E. Feyerabend was unimpressed. He pointed out that of course all the deductive consequences of E follow from E. This is trivial. But the original interest in reduction had been an interest in a relationship between actual theories in their historical context.[37]

[32] Feyerabend, 'On the "Meaning" of Scientific Terms', *J. Phil.* 62 (1965), 267–71; 'Consolations for the Specialist' in *Criticism and the Growth of Knowledge*, ed. I. Lakatos and A. Musgrave (Cambridge: Cambridge University Press, 1970), 220–1; 'Against Method: Outline of an Anarchistic Theory of Knowledge' in *Minnesota Studies in the Philosophy of Science* 4, ed. M. Radner and S. Winokur (Minneapolis: University of Minnesota Press, 1970), 84.

[33] Feyerabend, 'On the "Meaning" of Scientific Terms', 271–2.

[34] Feyerabend, 'Explanation, Reduction, and Empiricism', 76–81.

[35] David Hull, 'Informal Aspects of Theory Reduction' in *Boston Studies in the Philosophy of Science* 32, ed. R. S. Cohen *et al.* (Dordrecht: D. Reidel, 1976), 653–69.

[36] Hilary Putnam, 'How Not to Talk About Meaning' in *Boston Studies in the Philosophy of Science* 2, ed. R. Cohen and M. Wartofsky (New York: Humanities Press, 1965), 206–7.

[37] Feyerabend, 'Reply to Criticism: Comments on Smart, Sellers and Putnam' in *Boston Studies* 2, 229–30.

Theory replacement and the Correspondence Principle

Whewell's Tributary–River Model of scientific progress may be correct even though successive scientific theories do not satisfy Nagel's conditions for reduction. One way in which this might be the case is if successive theories within *HS* conform instead to the requirements of Niels Bohr's Correspondence Principle.

The Correspondence Principle is an axiom of Bohr's early quantum theory of the hydrogen atom (1913). It stipulates that the hydrogen electron obeys the laws of classical electrodynamics in the limit as the radius of its orbit approaches infinity. The Correspondence Principle ensures that the equations of motion for discrete electron orbits pass over into the classical equations for the case of an electron no longer bound to the nucleus of the atom.[38]

Bohr suggested that the Correspondence Principle is more than a principle of a specific theory. It is, in addition, a methodological directive. As such, it directs the theorist to build upon present theories in such a way that the successor theory and its predecessor are in agreement for that domain in which the earlier theory had been successful. The Correspondence Principle requires of each candidate to succeed theory *T* that

(1) the new theory have greater testable content than *T*, and
(2) the new theory be in asymptotic agreement with *T* in the region in which *T* is well confirmed.

Both Schrödinger's Quantum Theory and Einstein's Special Relativity Theory satisfy the Correspondence Principle. In the limiting case in which the quantum of action may be neglected, Schrödinger's formulae for the probability distribution of quantum-mechanical particles reduce to the classical equations of motion. And in the limiting case in which the velocity of a system is negligible with respect to the velocity of light, Einstein's formulae also reduce to the classical equations of motion.

Ernest Hutten pointed out that in neither case is classical mechanics a deductive consequence of the more inclusive theory. Nagel's conditions for reduction are not fulfilled. But to satisfy the Correspondence Principle it suffices that, under specified limiting conditions, calculations made from the equations of Special

[38] Niels Bohr, 'Atomic theory and Mechanics (1925) in *Atomic Theory and the Description of Nature* (Cambridge: Cambridge University Press, 1961), 35–9.

Relativity Theory approach asymptotically calculations made from the equations of classical mechanics.[39]

Feyerabend denied that the asymptotic agreement of mathematical formulae establishes the incorporation required by the Tributary–River Model of scientific progress. Indeed he insisted that *HS* reveals that successful high-level theories do *not* incorporate their predecessors. His conclusion is that

what happens when transition is made from a restricted theory *T'* to a wider theory *T* (which is capable of covering all the phenomena which have been covered by *T'*) is something much more radical than incorporation of the *unchanged* theory *T'* into the wider context of *T*. What happens is rather a complete replacement of the ontology of *T'* by the ontology of *T*, and a corresponding change in the meanings of all descriptive terms of *T'* (provided these terms are still employed).[40]

Although Feyerabend initially did not specify explicitly what counts as a 'change of meaning', he did indicate that the meaning of a term is a function of the theory in which it occurs.[41]

To his critics it appeared that Feyerabend was committed to the position that any change in the structure of a theory is also a change in meaning of the terms of the theory. Dudley Shapere complained that an alternative axiomatization ought not count as a change in the meanings of the terms of a theory.[42] Peter Achinstein emphasized that it would be stretching the phrase 'change of meaning' beyond all usefulness if every alteration of a theory were to count as a 'change of meaning' of its terms. Achinstein suggested a number of cases of theory modification that presumably do not change the meanings of the terms involved. His best illustration is perhaps the alteration of the Bohr Theory of the Hydrogen Atom to permit the electron to describe elliptical orbits around the nucleus. Surely the inclusion of elliptical orbits among those orbits permitted the electron does not alter the meaning of 'electron'.[43]

---

[39] Ernest Hutten, *The Language of Modern Physics* (New York: Macmillan, 1956), 166–7.
[40] Feyerabend, 'Explanation, Reduction, and Empricism', 59.
[41] Feyerabend, 'Problems of Empiricism' in *Beyond the Edge of Certainty*, ed. R. Colodny (Englewood Cliffs: Prentice–Hall, 1965), 180.
[42] Dudley Shapere, 'Meaning and Scientific Change' in *Mind and Cosmos*, ed. R. Colodny (Pittsburgh: University of Pittsburgh Press, 1966), 55–6.
[43] Peter Achinstein, 'On the Meaning of Scientific Terms', *J. Phil.* 61 (1964), 504–5.

In reply to Achinstein, Feyerabend conceded that not every change in theory produces a change of meanings. For instance, given T = classical celestial mechanics, and $\bar{T}$ = a theory of the same form with a slightly changed value of the strength of the gravitation potential, the transition from T to $\bar{T}$ does not involve any changes in meanings. Feyerabend noted that although T and $\bar{T}$ assign different force-values in a given application, the difference in values is *not* due to the action of different kinds of entities.[44] He contrasted this 'transition' with a transition from T to T', where T' is general relativity theory. In this latter transition, the meaning of 'spatial interval' does change, supposedly because the entities referred to differ in T and T'.

Table 3. Feyerabend's Criterion of Change of Meaning

Transition	Type of change in theory	Change of meaning of the terms of T_1 that occur in T_2
From T_1	1. Alteration of the system of classes to which the concepts of T_1 refer.	Yes
To T_2	2. Alteration of the extensions of the classes, but not the way in which the classes are circumscribed.	No

Consistent with this contrast, Feyerabend suggested the criterion of 'change of meaning' expressed in Table 3.

Of course, this criterion of meaning-change is useful only if unique and definite rules of classification can be specified for the 'entities' referred to by theories. Shapere emphasized the difficulties that arise upon application of the criterion. The rules must be sufficiently definite to permit an unambiguous classification. And if competing rules of classification are available, then it must be clear which rule is used implicitly by the theory in question. Shapere expressed doubt that this can be achieved in the case of high-level theories.

Are mesons different 'kinds of entities' from electrons and protons, or are they simply a different subclass of elementary particles? Are the light rays of classical mechanics and of general relativity (two theories which Feyerabend claims are 'incommensurable') different 'kinds of entities' or

[44] Feyerabend, 'Comments and Criticisms on the "Meaning" of Scientific Terms', *J. Phil.* 62 (1965), 267.

not? Such questions can be answered *either* way, depending on the kind of information that is being requested . . . for there are differences as well as similarities between electrons and mesons, as between light rays in classical mechanics and light rays in general relativity.[45]

Achinstein raised a question about the comparability of theories. Feyerabend had suggested that the meanings of the descriptive terms of successive high-level theories change. Achinstein pointed out that if no descriptive term in T has the same meaning as any term in T', then T cannot contradict T'. But in *HS* one high-level theory often denies what is asserted by some other high-level theory. For example, in the Bohr Theory the angular momentum of an electron is quantized, and this contradicts the classical electromagnetic theory.[46]

Feyerabend replied that low-level theories may be compared by comparing their consequences with 'what is observed'. This can be done because there exist background theories to provide a common interpretation for the observational consequences of low-level theories. The background theories have a status that is independent of the fate of individual low-level theories. But high-level theories cannot be compared in this way. No theory-neutral observation language is available at this level. Rather, each high-level theory specifies its own observation language.

Nevertheless, Feyerabend suggested that the observational consequences of high-level theories may be compared with 'human experience as an actually existing process'.[47] Feyerabend recommended a 'pragmatic theory of observation', according to which which

a statement will be regarded as observational because of the *causal context* in which it is being uttered, and *not* because of what it means. According to this theory, 'this is red' is an observation sentence, because a well-conditioned individual who is prompted in the appropriate manner in front of an object that has certain physical properties will respond without hesitation with 'this is red'; and this response will occur independently of the *interpretation* he may connect with the statement.[48]

An observer is *caused* to respond in certain ways by the characteristics of the observational situation and his prior conditioning. Feyerabend declared that 'we can . . . determine in a

[45] Shapere, 'Meaning and Scientific Change', 64.
[46] Achinstein, 'On the Meaning of Scientific Terms', 499.
[47] Feyerabend, 'Problems of Empiricism', 214.   [48] Ibid., 198.

straightforward manner whether a certain movement of the human organism is correlated with an external event and can therefore be regarded as an indicator of this event'.[49] Feyerabend proposed to use these verbal 'event-indicators' to evaluate high-level theories. A statement that a given 'event-indicator' took place may be consistent with one theory and inconsistent with a second theory.

Shapere complained that, although Feyerabend denied that there could exist a theory-independent observation language, he made use of theory-independent observations to evaluate theories.[50] But Shapere denied that these theory-independent observations could perform the function that Feyerabend assigned to them. Even if observation sentences do issue from certain situations as conditioned responses of the observer, they are mere uninterpreted noises. They have no more linguistic content than a burp.[51] Before such a sentence can count as a test of a theory, it must be interpreted. Feyerabend had maintained, however, that observational findings are subject to reinterpretation at the hands of successful new theories. If this is so, then observation sentences cannot decide the issue between competing theories. In short, Feyerabend cannot have it both ways. *Either* an appeal to observation sentences can decide the issue between competing theories and observation sentences are not subject to reinterpretation by the theories compared, *or* an appeal to observation sentences cannot decide the issue between competing theories and observation sentences are subject to reinterpretation by the theories compared.

Of course, one can *decide* to reject a theory upon the occurrence within experience of a conflict between expectations derived from the theory and an observer's response to an observational situation. This is to decide not to interpret the response from the standpoint of the theory. But Feyerabend can select this option only at the risk of introducing disharmony into his methodology, since he has repeatedly emphasized the importance of theoretical reinterpretation of statements about what is observed.

Supporters and critics of the Tributary–River View have appealed to *HS* to support their respective positions on the nature of scientific progress. Their arguments about Nagelian Reduction and applications of the Correspondence Principle have depended critically upon interpretations of developments in science.

[49] Ibid., 212.
[50] Shapere, 'Meaning and Scientific Change', 60.   [51] Ibid., 60.

4

Prescriptive Philosophy of Science:
A Historical Survey

IF the analysis of the preceding chapters is correct, then *PS* and *HS* are not mutually exclusive disciplines. The nature and extent of the relationship between them remains to be specified. Much depends on what is understood by 'philosophy of science'.

It will be assumed in this study that *PS* is a normative discipline. Its principal concern is the formulation of evaluative *standards*. Specific instances of evaluative practice in science are relevant to *PS* only in so far as they are a source or a warrant of claims about norms. The phrase 'normative status' is systematically ambiguous, however. 'Normative status' may be ascribed to *PS* in a prescriptive sense. A prescriptive *PS* specifies, and seeks to justify, standards by which scientific hypotheses, theories, and explanatory arguments *ought* to be evaluated. The prescriptivist philosopher of science sets forth and recommends a set of evaluative standards, application of which supposedly contributes to the creation of 'good science'. Alternatively, 'normative status' may be ascribed to *PS* in a descriptive sense. Descriptive *PS* formulates evaluative standards that explicitly or implicitly have informed scientific developments. No recommendations about 'proper' evaluative practice are issued within descriptive *PS*.

Consider the descriptive version of *PS*. That *PS* is both normative and descriptive may be taken as either an empirical claim about the nature of *PS* or a stipulative proposal for a reinterpretation of the purpose and scope of *PS*. The empirical claim is false. To take *PS* to be nothing but a comparative study of evaluative practices is to ignore an extensive tradition of prescriptive *PS*.

The following historical survey is a selective, but perhaps sufficient, indication of widespread agreement about the prescriptive nature of the discipline. Each of the philosophers of science whose work is discussed did accord prescriptive significance to *PS*. They did so by recommending, and seeking to justify, evaluative standards which they held to be important to the creation of good science.

I ARISTOTLE

Aristotle did not discuss explicitly the distinction between prescriptive and descriptive analyses of scientific method. His comments about scientific method indicate, however, that he took what we now call 'philosophy of science' to be a prescriptive discipline. Aristotle recommended standards by which proposed scientific interpretations *ought* to be evaluated, claimed that certain applications of these standards are correct, and advanced justificatory arguments on behalf of the standards. By so doing, he formulated a prescriptive *PS*.

The most important of Aristotle's normative recommendations is a theory of scientific procedure in which the scientist progresses from a knowledge of facts to an understanding of the causes of the facts being what they are. The procedure consists of an inductive stage which progresses from observation reports to explanatory principles, and a deductive stage in which a statement about the observation reports is exhibited as the conclusion of premises that include explanatory principles.

Aristotle held that a successful application of the inductive–deductive procedure culminates in a *demonstration* in which the conclusion follows logically from the stated premises and the attribution asserted in the conclusion could not be other than it is. He declared that

we suppose ourselves to possess unqualified scientific knowledge of a thing, as opposed to knowing it in the accidental way in which the sophist knows, when we think that we know the cause on which the fact depends, as the cause of that fact and of no other, and, further, that the fact could not be other than it is.[1]

Aristotle distinguished carefully between demonstrations and 'merely accidental' deductive arguments in which the conclusion follows logically from true premises, but which premises do not state the cause of the attribution made in the conclusion.

Aristotle cited examples of this contrast. One such example is the pair of arguments below:

---

[1] Aristotle, *Posterior Analytics*, 71^b 8–11.

(1) All bodies near the earth are bodies that shine steadily.
　　　All planets are bodies near the earth.
　　∴ All planets are bodies that shine steadily.

(2) All bodies that shine steadily are bodies near the earth.
　　　All planets are bodies that shine steadily.
　　∴ All planets are bodies near the earth.[2]

Syllogism (1) demonstrates its conclusion, syllogism (2) does not.

Aristotle recognized the need to specify appropriate criteria of causal relatedness. He maintained that in a causal relation that which is attributed to a subject is true of every instance of the subject, is true of the subject precisely and not in virtue of a relationship to a more inclusive subject, and is true 'essentially' of the subject. Aristotle's criteria are of limited usefulness. His intent is clear, however. He sought evaluative criteria, application of which would lead to the creation of 'good science'.

Aristotle also advanced justificatory arguments for several of the evaluative standards that he recommended. He argued, for instance, that the first-figure syllogism is the most appropriate figure for the deductive stage of scientific enquiry because it is the only figure that 'enables us to pursue knowledge of the essence of a thing'.[3] He noted that neither the second figure nor the fourth figure permit an affirmative conclusion. Hence they cannot reveal the essential characteristics of a thing. He noted also that, since essentiality implies universality, the third figure also cannot reveal the essential characteristics of a thing.

Aristotle also sought to justify inclusion of an inductive stage of enquiry in his theory of scientific procedure. He observed that two groups of critics deny the necessity of an inductive stage. One group does so because its members believe that all truths are demonstrable. Aristotle dismissed this position because an infinite regress of demonstrations of premisses cannot be completed and no circular demonstration is possible.

According to Aristotle, a second group of critics denies the necessity of an inductive stage of scientific enquiry on the grounds that there is no such thing as genuine scientific knowledge. It is the position of the members of this second group that, since the most general premisses cannot themselves be demonstrated, all that can be achieved in science is an unpacking of the logical consequences

---

[2] Ibid., 78^a 30–78^b 2.　　　[3] Ibid., 79^a 23–4.

of a set of suppositions. Aristotle pointed out that this position is based on the assumed premiss that all scientific knowledge is demonstrative knowledge. He maintained that this assumed premiss is false because truths which are both universal and necessary may be developed by inductive inference from reports of the sensory perception of particulars. He declared that

when one of a number of logically indiscriminable particulars has made a stand, the earliest universal is present in the soul: for though the act of sense-perception is of the particular, its content is universal—is man, for example, not the man Callias. A fresh stand is made among these rudimentary universals, and the process does not cease until the indivisible concepts, the true universals, are established.[4]

Clearly, Aristotle both recorded evaluative standards and advanced arguments in support of these standards. His *PS* is resolutely prescriptive.

II NEWTON

Aristotle had viewed scientific enquiry as a progression from knowledge of facts to knowledge of the reasons why the facts are as they are. On this view, a scientific investigation is complete once a demonstration of the initial facts has been achieved.

Isaac Newton augmented the Aristotelian theory of procedure by emphasizing the experimental confirmation of deductive consequences which go beyond the original inductive evidence. In letters to Pardies and Linus, Newton referred to his two-prism experiment as an 'Experimentum Crucis'.[5] The two-prism experiment was designed to test the 'theory' that sunlight comprises rays of different colours, each of which has a characteristic degree of refrangibility. Newton deduced from the theory the consequence that light of a specific colour, isolated after refraction of sunlight by a prism and directed upon a second prism, should be refracted by the angle characteristic of that colour without resolution of the beam into other colours.[6]

Newton was not the first to emphasize the experimental

---

[4] Ibid., 100^b 14–21.
[5] I. Bernard Cohen (ed.), *Isaac Newton's Papers and Letters on Natural Philosophy* (Cambridge, Mass.: Harvard University Press, 1958), 91, 107, 154, 159–60.
[6] Isaac Newton, *Opticks* (New York: Dover, 1952), 45–8.

confirmation of deductive consequences that go beyond the initial evidential base for a theory. The criterion had been recommended by Francis Bacon and Galileo. However, the experimental confirmation of 'new' consequences gained status as an important evaluative standard largely because of the influence of Newton's *Opticks*.

Newton sought to convert his fellow natural philosophers to a particular vision of the discipline. Henceforth natural philosophy was to be 'Experimental Philosophy', the content of which is restricted to statements about 'manifest qualities' (those aspects of phenomena whose values can be measured experimentally), 'theories' which state relations among manifest qualities, and queries directive of further enquiry.

When Newton claimed 'I feign no hypotheses,'[7] he meant this to be taken as a covert directive principle. Within Experimental Philosophy it is inappropriate to put forward as explanatory premises hypotheses about 'occult qualities' whose values are not subject to experimental determination. Hypotheses may have heuristic value as queries that suggest fruitful programmes of research, but hypotheses are not to be used as if they had the same assured status as theories.[8]

Despite his aversion to hypotheses, Newton himself speculated about underlying mechanisms that might account for observed relations among manifest qualities. In particular, he entertained a hypothesis about an ethereal medium responsible for gravitational attraction and a hypothesis about an interaction of corpuscles responsible for the reflection and refraction of light. Newton was usually careful to indicate that these hypotheses were to be taken as queries directive of future enquiry.

Newton hoped that the practice of Experimental Philosophy would consolidate and extend the success achieved by the theory of universal gravitation. By focusing on such manifest qualities as force, position, and velocity, practitioners of Experimental Philos-

---

[7] Newton, *Mathematical Principles of Natural Philosophy*, trans. A. Motte, rev. F. Cajori (Berkeley: University of California Press, 1962), 2. 547.

[8] Newton's use of the terms 'theory' and 'hypothesis' does not conform to modern usage. Theories state invariant relations among manifest qualities. The laws of refraction and gravitational attraction are 'theories' in Newton's usage. Hypotheses, by contrast, are speculations about underlying mechanisms for which no experimental techniques of measurement are available.

ophy were to develop theories of electrical, magnetic, and chemical interaction. He declared that

> to tell us that every Species of Things is endow'd with an occult specifick Quality by which it acts and produces manifest Effects, is to tell us nothing: But to derive two or three general Principles of Motion from Phaenomena, and afterwards to tell us how the Properties and Actions of all corporeal Things follow from those manifest Principles, would be a very great step in Philosophy, though the Causes of those Principles were not yet discover'd.[9]

To facilitate this research programme, Newton suggested a set of four directive principles or 'rules of reasoning in philosophy':

I. We are to admit no more causes of natural things than such as are both true and sufficient to explain their appearances . . .

II. Therefore to the same natural effects we must, as far as possible, assign the same causes . . .

III. The qualities of bodies, which admit neither intensification nor remission of degrees, and which are found to belong to all bodies within the reach of our experiments, are to be esteemed the universal qualities of all bodies whatsoever . . .

IV. In experimental philosophy we are to look upon propositions inferred by general induction from phenomena as accurately or very nearly true, notwithstanding any contrary hypotheses that may be imagined, till such time as other phenomena occur, by which they may either be made more accurate, or liable to exceptions.[10]

Newton thus issued sweeping prescriptive pronouncements about proper method in science. Many eighteenth-century scientists embraced the programme of Experimental Philosophy.[11]

Newton perceived that the chief rival position was that of Cartesian Philosophy, a basic assumption of which is that motion can be produced only by impact or pressure. He advanced justificatory arguments of two types in support of Experimental Philosophy. One was to contrast his own theory of the solar system, the pre-eminent achievement of Experimental Philosophy,

---

[9] Newton, *Opticks*, 401–2.

[10] Newton, *Mathematical Principles*, 2. 398–400.

[11] See, for instance, Arnold Thackray, *Atoms and Powers* (Cambridge: Harvard University Press, 1970); I. Bernard Cohen, *Franklin and Newton* (Cambridge, Mass.: Harvard University Press, 1966).

with the Cartesian account based on the hypothesis that planets are carried around the sun by a vortex of invisible particles of ether. He proved that the Cartesian Vortex Hypothesis has consequences inconsistent with Kepler's Laws of Planetary Motion.

However, Newton was not content to rest the case for the superiority of Experimental Philosophy on the question of consistency with Kepler's Laws. He believed that there were as well good theological reasons to prefer Experimental Philosophy. In the General Scholium at the end of Book III of the *Principia*, Newton inserted a discussion of the Divine Nature.[12] He observed that it is insufficient to characterize God as 'eternal', 'omnipotent', 'omniscient', and 'perfect'. Of course, it is true to predicate these 'absolute terms' of God. But, in addition, it is most important to emphasize God's Lordship over creation. The term 'Lord' is a relational term which emphasizes the continuing dependence of God's creatures upon their Creator.

This reference to God's 'Lordship' over creation is not merely an aside. Newton believed that it is the task of the natural philosopher to discover how the Creator manifests his presence within the universe. There occur important references to God as Governor of the universe in the *Principia*, the *Opticks*, and various of Newton's papers.

Newton maintained that there were good *scientific* reasons to believe that God's relation to the universe was that of Creator *and* Governor. Were it not for God's continual sustaining activity, the solar system would become unstable. Newton pointed out two sources of potential instability which, if not counteracted, would prove fatal to the solar system.

The first source of potential instability is the passage through the solar system of a large number of comets. In their repeated forays through the system, comets pass close to the several planets and their satellites. Despite the disturbing influence of these visitors, the planets continue to describe roughly elliptical orbits around the sun. 'This most beautiful system of the sun, planets, and comets', he wrote, 'could only proceed from the counsel and dominion of an intelligent and powerful Being'.[13]

A second source of potential instability is the continual loss of 'motion' in the solar system. Newton sometimes spoke of a

---

[12] Newton, *Mathematical Principles*, 2, 544–6.
[13] Ibid., 544.

tenuous ether through which the planets move in their circuits around the sun.[14] If the planets do indeed move through such a medium, some motion is bound to be lost. Newton noted that collisions between bodies that are not perfectly elastic result in a net loss of momentum. But since the solar system is not winding down, God must intervene to restore lost motion. Newton attributed the continuing regular motion of the planets—in the face of this resistance—to the Divine Governance of the universe.[15]

Newton was also puzzled by anomalies in the motions of Jupiter and Saturn. The observed motions of these two planets deviate considerably from elliptical orbits. Newton was unaware that the irregularities in the motions of Jupiter and Saturn are cyclical and he took the stability of the solar system in the face of these anomalies to be yet another indication of the Divine Governance of the system.

Newton found the Cartesian view of the universe repugnant on theological grounds. On the Cartesian view, after God's initial act of creation, every motion is caused by contact with other bodies. Newton protested that God's relation to the universe is one of continuing sustenance and governance, and he accused the Cartesians of excluding the Divine Will from the operation of the universe. In Newton's universe God did not retire to the sidelines after the initial act of creation.

Experimental Philosophy is consistent with God's governance of the universe. The Cartesian view, by contrast, promotes atheism. The Cartesian scientist may practise his craft exclusively within the scope of materialistic assumptions which require reference to God only for an initial act of creation.

This indictment may not have been wholly fair, but there is no doubt that Newton took it quite seriously. By recommending and seeking to justify a methodological orientation for science, Newton exhibited a commitment to a prescriptive *PS*.

[14] Cf. *Opticks*, Query 31. But in the General Scholium to Book III of the *Principia*, he spoke of the space between the planets as void of all matter.

This indecision is reflected in Newton's multiple positions on the 'cause' of gravitational attraction. Newton defended, at various times, three different positions. The first position, and the one to which he retreated when challenged, was an agnostic confession that he had no idea what caused gravitational attraction. The second was that there exists in the solar system a tenuous ether which transmits by contact a force on the planets. And the third was that the cause of gravitational attraction is a 'nonmechanical force' that produces motion without impact.

[15] Newton, *Opticks*, 398–9.

III NINETEENTH-CENTURY METHODOLOGISTS: HERSCHEL, WHEWELL, AND MILL

Francis Bacon had singled out various types of 'prerogative instances' which are of special value in the search for scientific laws. The most famous of these is the 'instance of the fingerpost',[16] evidence which provides inductive support for one interpretation and refutation for its competitors. Having arrived at an intersection, the scientist who has knowledge of an 'instance of the fingerpost' knows which path to travel.

John Herschel included the instance of the fingerpost—renamed the 'crucial experiment'—among criteria of acceptability for scientific theories. By so doing he removed discussion of this type of instance from a Baconian context of discovery to a context of justification. Herschel emphasized that some types of confirmation provide greater evidential support than others. Particularly important are: (1) the crucial experiment, (2) the uncovery of undesigned scope,[17] (3) the successful application of a law to extreme cases, and (4) the favourable redetermination of initially recalcitrant data.[18]

In addition, Herschel recommended a number of rules to guide research into causal connections. Among these rules are the methods of agreement, difference, concomitant variations, and residues subsequently championed by John Stuart Mill.[19]

Herschel sought to justify the evaluative standards he had formulated by showing that the development of allegedly progressive episodes within *HS* exemplifies the application of these standards. For instance, the experimental determination of the identical acceleration of a coin and a feather in a vacuum exemplifies the successful application of a law to extreme cases;[20] Galileo's telescopic evidence of Venus' full range of phases exemplifies the crucial experiment (heliocentrism confirmed, Ptolemaic geocentrism refuted); and Laplace's application of the theory of heat to explain a discrepancy between calculated and observed velocities of sound exemplifies the uncovery of undesigned scope.[21]

---

[16] Francis Bacon, *Novum Organum* 2, Aphorism XXXVI.

[17] Herschel's position on 'undesigned scope' is discussed in Chapter 3.

[18] John F. W. Herschel, *A Preliminary Discourse on the Study of Natural Philosophy* (London: Longman; etc., 1830), 165–88.

[19] Ibid., 152–8.   [20] Ibid., 168.   [21] Ibid., 171–2.

For the most part, Herschel was content to specify an episode or two that conforms to the requirements set by a particular evaluative standard. He was not clear on the relationship between the fact that a historical episode exemplifies application of an evaluative standard and the justification of that standard. However, it is clear that he believed that such exemplification is important and that scientists ought to apply the evaluative standards thus exemplified.

Whewell and Mill shared Herschel's commitment to a prescriptive *PS*.[22] Whewell recommended consilience of inductions as a criterion of acceptability of scientific theories and sought to justify the criterion by an appeal to *HS*.[23] Mill recommended a method of difference, proper instantiation of which proves causal connection, and sought to justify this schema by appeal to a principle of uniformity itself established by simple enumeration.[24]

IV   CAMPBELL

In *Physics: The Elements*[25] (1920), N. R. Campbell set forth an incisive analysis of the structure of scientific theories. According to Campbell, a scientific theory is an interpreted axiom system with both deductive links and analogical links to experimental laws. He maintained that there are three necessary, and jointly sufficient, conditions of the adequacy of a scientific theory. First, a formal condition must be satisfied. A theory contains a self-consistent axiom system (a 'hypothesis') and rules of correspondence (a 'dictionary') which link some, but not necessarily all, non-logical terms of the axiom system to empirically determinable magnitudes. Second, the interpreted axiom system must imply one or more experimental laws. And third, 'the propositions of the hypothesis must be analogous to some known laws'.[26] Campbell insisted that it is only in virtue of an analogy to a system governed by previously established laws that a theory *explains* the laws which are its deductive consequences.

---

[22] The positions of Whewell and Mill are discussed in Chapter 6.
[23] William Whewell, *Novum Organon Renovatum* (London: John W. Parker & Son, 1858), 90.
[24] John Stuart Mill, *System of Logic* (London: Longmans, Green, 1865), 2. 101.
[25] Reissued as *Foundations of Science* (Dover, 1957).
[26] Norman R. Campbell, *Foundations of Science*, formerly *Physics: The Elements* (New York: Dover, 1957), 129.

Campbell was aware that the requirement that a theory display an analogy is controversial. To show the necessity of this requirement he created an interpreted axiom system which implies the experimental law that the electrical resistance of a piece of pure metal is directly proportional to its absolute temperature.[27]

Campbell's interpreted axiom system implies a well-confirmed experimental law. But as he pointed out, 'any fool can invent a logically satisfactory theory to explain any law'.[28] What is lacking in the case at hand is that the interpreted axiom system 'does not display any analogy; it is just because an analogy has not been used in its development that it is so completely valueless'.[29]

Some scientific theories incorporate analogies to 'mechanical' systems. In the kinetic theory of gases the analogy is to colliding miniature billiard balls. For many theories, however, the analogy is one of mathematical form. Campbell called attention to Fourier's theory of heat conduction for which the 'hypothesis' is

$$\lambda \left(\frac{\partial^2 \theta}{\partial x^2} + \frac{\partial^2 \theta}{\partial y^2} + \frac{\partial^2 \theta}{\partial z^2} \right) = \rho c \frac{\partial \theta}{\partial t}$$

and the 'dictionary' specifies that θ is the absolute temperature, λ is the thermal conductivity, ρ is the density, c is the specific heat, t is the time, and x, y, and z are the spatial coordinates of a point in an infinite plane parallel slab.[30] The analogy in this theory is between the mathematical form of the hypothesis and the mathematical form of various experimental laws for finite slabs of specific materials. Of course these experimental laws are precisely the laws that are implied by the hypothesis-plus-dictionary. Campbell emphasized that theories of this type also discharge their explanatory function only by displaying an analogy. By formulating

---

[27] The hypothesis consists of the following mathematical propositions:

(1) u, v, w, . . . are independent variables.
(2) a is a constant for all values of these variables.
(3) b is a constant for all values of these variables.
(4) $c = d$, where c and d are dependent variables.

The dictionary consists of the following propositions:

(1) The assertion that $(c^2 + d^2)\, a = R$ where R is a positive and rational number, implies the assertion that the (electrical) resistance of some definite piece of pure metal is R.

(2) The assertion that $\frac{cd}{b} = T$ implies that the (absolute) temperature of the same piece of pure metal is T (Ibid., 123).

[28] Ibid., 129.   [29] Ibid., 130.   [30] Ibid., 140–1.

and defending criteria by which scientific theories ought be evaluated Campbell contributed to the tradition of prescriptive *PS*.

V OPERATIONALISM

P. W. Bridgman's principal contribution to *PS* is a normative pronouncement about the proper relationship between scientific concepts and instrumental procedures. He insisted that every concept not linked to measuring procedures be excluded from science.[31] Bridgman made explicit a methodological position already accepted by certain scientists—notably Mach, Poincaré, Duhem, and Einstein.

Mach, for instance, in his reformulation of Newtonian mechanics, suggested a definition of 'mass' in terms of the results of operations performed. The definition stipulated, in part, that the ratio of two masses is equal to the inverse ratio of the accelerations of the two bodies, observed under specified conditions. Mach stressed that a definition of 'mass' in terms of observed motions is clearly superior to any purely verbal definition in terms of 'quantity of matter'.[32] Mach also examined Newton's concepts of Absolute Space and Absolute Time, and concluded that, since no operations can be performed to assign values to these concepts, they should be eliminated from physics.[33]

Poincaré, adopting a similar standpoint, stated the general principle that a concept is useful in science only if we know how to measure its values. For this reason, he criticized the claim that the concept of force is an extrapolation from our direct intuitive apprehension of muscular exertion. What counts, according to Poincaré, is not what force is 'in itself', but rather knowledge of how to measure it.[34]

Duhem applied this operational requirement to scientific theories. He maintained that a theory can be empirically significant only if its conclusions make assertions about concepts whose values can be measured. But he also recognized that not every term of a

[31] P. W. Bridgman, *The Logic of Modern Physics* (New York: Macmillan, 1960), 28–9; *The Nature of Physical Theory* (New York: Dover, 1936), 9–12.

[32] Ernst Mach, *The Science of Mechanics* (LaSalle: Open Court, 1960), 265–71.

[33] Ibid., 271–97.

[34] Henri Poincaré, *Science and Hypothesis*, trans. G. B. Halsted (New York: The Science Press, 1905), 73–8.

measurements of strains on the surface of the body. The value of theory need be linked to measuring operations. Duhem stipulated that the use of intervening variables is legitimate provided that the deductive consequences of the theory in which these variables are embedded are confirmed by experience.[35]

It was Einstein's discussion of the concept of simultaneity, however, which most impressed Bridgman. Einstein had challenged the widespread assumption that simultaneity is an objective relation between events. By raising questions about operations required to judge two events simultaneous, he showed that simultaneity is a relation between two or more events and an observer.

Bridgman recommended a generalization of Einstein's methodological approach. It is instrumental operations by which values are assigned that give significance to a scientific concept. There are no such operations for 'absolute simultaneity' and this concept is to be excluded from science. A similar fate is appropriate for 'the simultaneous position and momentum of a subatomic particle'.

Bridgman included among permissible operations, 'paper and pencil' operations.[36] He acknowledged that the relation between values assigned to a scientific concept and measuring procedures may be complex, but he insisted that there must be some such relationship. Scientific concepts can be graded, roughly at least, on the extensiveness of 'paper and pencil' operations that intercede between instrumental operations and the assignment of values to the concept. Consider the following list:

'local length'
'stress within a deformed body'
'entropy change for an irreversible process'
'velocity of an individual gas molecule'
'Ψ-function'

Values of 'local length' are determined by laying a properly calibrated measuring rod on the object to be measured, noting the points on the scale which correspond to the two ends of the object, and taking its length to be the difference between these two points. Values of 'stress' are computed via mathematical theory from

---

[35] Pierre Duhem, *The Aim and Structure of Physical theory* (2nd edn., 1914), trans. P. Wiener (New York: Atheneum, 1962), 19–23.

[36] Bridgman, *Reflections of a Physicist* (New York: Philosophical Library, 1950), 15.

an 'entropy change in an irreversible process' (for example Joule–Thomson expansion into a vacuum) is calculated by devising a reversible path between the same initial and final states and measuring the heat absorbed in the reversible process. The value of 'an individual molecular velocity' cannot be determined within the kinetic theory of gases. However, the root-mean-square velocity of all the molecules may be determined by measuring the temperature of the gas. Apart from kinetic theory, individual molecular velocities may be measured by a molecular-beam apparatus. Values of the 'Ψ-function', by contrast, involve the term '$e^{\sqrt{-1}}$', and, *in principle* cannot be measured instrumentally. Nevertheless the Born Interpretation correlates $\mid \Psi \mid^2$, but not Ψ itself, with instrumentally determinable values of electron charge densities, transition frequencies, and scattering distributions. The link between Ψ and the results of instrumental operations is indirect and complex. But there is a link, and this is sufficient to qualify the Ψ-function as an operationally significant concept.

Bridgman claimed that the operational criterion of empirical meaningfulness is an epistemological lesson to be drawn from Einstein's analyses in the Theory of Special Relativity. A justificatory argument in support of the operational criterion is implicit in Bridgman's position, namely:

> Any methodological principle implicated in the formulation of a successful high-level theory is a principle that ought be promulgated as a general methodological principle. The operational criterion is implicated in Special Relativity Theory. Hence the operational criterion ought be promulgated as a general methodological principle.

VI   THE PROGRAMME OF THE VIENNA CIRCLE

From its inception in the early 1920s, the Vienna Circle[37] staked out a normative programme for *PS*. Carnap, Neurath, and Hahn issued a manifesto in 1929 which outlined a two-part programme of reform. The first task was to purge science of metaphysical speculation. Members of the Circle believed that the exclusion from science of claims about 'vital forces', 'phlogiston', the

---

[37] Participants in the early discussions of the Vienna Circle included Rudolf Carnap, Herbert Feigl, Philipp Frank, Kurt Gödel, Hans Hahn, Otto Neurath, and Friedrich Waismann.

'luminiferous ether', and the like, had contributed to progress in science. What was needed, they felt, was a rigorous logical analysis of language that would show that metaphysical claims lack empirical significance. This analysis could then be used to insure that futher metaphysical conceits do not arise within science. The second task was to provide a secure epistemological foundation for the sciences.

Members of the Vienna Circle emphasized that the evaluation of proposed scientific interpretations is a two-step procedure. The first step is to ascertain whether the interpretation is empirically significant. The second step is to assess the acceptability of those interpretations that do qualify as empirically·significant.

Circumscription of the range of empirically meaningful interpretations proved to be a difficult problem. Various criteria of empirical significance were proposed—verifiability, falsifiability, confirmability, translatability into an empiricist language—only to be abandoned as too inclusive or too exclusive.[38] Nevertheless, the search for an appropriate criterion of empirical meaningfulness was undertaken from a prescriptive perspective. Its intent was to prescribe proper procedure to those who would practise science.

The second task acknowledged in the Vienna Circle Manifesto was to provide a firm empirical anchor for scientific theories. By and large, members of the Circle agreed that this was a matter of specifying the ways in which the theoretical claims of the sciences are related to 'elementary propositions' whose truth or falsity can be determined directly and unambiguously.

How this might be achieved had been indicated by Ludwig Wittgenstein. In the *Tractatus Logico-Philosophicus* (1921),[39] Wittgenstein had outlined a method for formulating complex propositions from a set of logically independent elementary propositions. The resulting language was truth-functional throughout; given a set of elementary propositions, the truth value of each of the complex propositions that can be formed from them is determined. For example, if 'p' is true and 'q' is false, then 'p or q' is true and 'p and q' is false. Wittgenstein succeeded in extending his method to universally quantified propositions. These propositions

---

[38] See Carl G. Hempel, 'The Empiricist Criterion of Meaning' in *Logical Positivism*, ed. A. J. Ayer (Glencoe: Free Press, 1959), 108–29.

[39] Ludwig Wittgenstein, *Tractatus Logico-Philosophicus*, trans. D. F. Pears and B. F. McGuinness (New York: Humanities Press, 1961).

are also truth-functionally dependent on the elementary propositions. Thus if the truth values of the elementary propositions are known, the truth-values of both molecular propositions and universally quantified propositions can be ascertained by logical analysis.

But how can the truth-values of the elementary propositions themselves be determined? According to Wittgenstein, an elementary proposition (1) exhibits the logical form of a state of affairs, and (2) asserts the existence of this state of affairs.[40] Supposedly, one determines the truth or falsity of an elementary proposition by ascertaining whether or not the state of affairs asserted to exist does in fact exist.

Wittgenstein did not give a single example of an elementary proposition. But members of the Vienna Circle, who saw in the *Tractatus* important possibilities for the logical analysis of science, did seek to specify the nature of elementary propositions.

Two rival interpretations were advanced. One interpretation was that elementary propositions are introspective reports of states of conscious awareness. 'Here, now red' and 'Here, now sour taste' might qualify as elementary propositions on this interpretation. The second interpretation was that elementary propositions are reports of the intersubjectively observable properties of physical objects. 'This object is red' and 'That surface is smooth' might qualify as elementary propositions on this interpretation.

The objectivist interpretation gradually gained ascendancy. It was from this standpoint that Rudolf Carnap made a striking attempt to put flesh on the *Tractatus* skeleton.[41] Carnap sought to show how the concepts of the sciences could be constructed out of a basic empirical vocabulary of observation predicates such as 'red' and 'smooth'. He noted that some scientific concepts may be introduced by explicit definitions which state a logical equivalence between the concept and a defining phrase. An example given by Carnap is that 'x is an arthropod' is logically equivalent to 'x is an animal and x has a segmented body and x has jointed legs'.[42]

Carnap recognized, however, that dispositional terms such as 'soluble', 'pressure', and 'magnetic field strength' cannot be

[40] Ibid., 41.
[41] Rudolf Carnap, 'Testability and Meaning', *Phil. Sci.* 3 (1936) and 4 (1937). Reprinted in *Readings in the Philosophy of Science*, ed. H. Feigl and M. Brodbeck (New York: Appleton–Century–Crofts, 1953).
[42] Carnap, 'The Elimination of Metaphysics Through Logical Analysis of Language' in *Logical Positivism*, ed. A. J. Ayer (Glencoe: Free Press, 1959), 63.

defined in this manner. In the case of 'soluble', a proposed explicit definition might be

$$(x) [Sx \equiv (Px \supset Dx)]^{43}$$

This proposed definition is inadequate. It qualifies as 'soluble' any substance not now placed in a liquid.[44]

Carnap suggested that, although dispositional predicates like 'soluble' cannot be defined explicitly in terms of observation predicates, they can be introduced by means of 'reduction sentences' or contextual definitions.[45] In the case of 'soluble', a useful contextual definition is

$$(x) [Px \supset (Sx \equiv Dx)] \ .$$

A contextually defined term cannot be eliminated in favour of a logically equivalent defining phrase. The term defined occurs in the consequent of an 'if . . . then . . .' expression. Thus, even if the descriptions of operations performed and results achieved are given in the vocabulary of observation predicates, the dispositional term is given only a *partial* meaning in terms of 'directly observable' properties. The term is defined only with respect to a specified set of operations. No restriction is placed on the application of the term in circumstances in which these operations are not performed.

Carnap maintained in 1938 that the theoretical terms of the sciences could be introduced by contextual definitions in terms of the observable properties of bodies.[46] But further enquiry led him

---

[43] 'For all x, x is soluble in a liquid if, and only if, if x is placed in the liquid, then x dissolves.'

[44] The truth-table for the material implication relation ' $\supset$ ' is

p	q	$p \supset q$
T	T	T
T	F	F
F	T	T
F	F	T

For case a, if Pa is false, then $(Pa \supset Da)$ is true. If we require that a contextual definition be true in order to be an acceptable definition, then Sa must be true also.

[45] Carnap, 'Testability and Meaning' in *Readings in the Philosophy of Science*, 52–6.

[46] Carnap, 'Logical Foundations of the Unity of Science' in *International Encyclopedia of Unified Science, vol. I, no. I*, reprinted in *Readings in Philosophical Analysis*, ed. H. Feigl and W. Sellars (New York: Appleton–Century–Crofts, 1949), 416–18.

to withdraw this claim. He later acknowledged that certain very important scientific concepts cannot be introduced in this way.[47] Among these concepts are the 'velocity of an individual molecule' in the kinetic theory of gases and the 'Ψ-function' in quantum mechanics. These theoretical concepts are terms in axiom systems. Each axiom system as a whole is linked to observation predicates. But no contextual definitions are provided for the individual terms in question. It became clear from Carnap's work that statements about theoretical terms cannot be constructed out of equivalent expressions using the observational vocabulary. The image of an empiricist language constructed out of a primitive vocabulary of observation predicates is not an image that can be superimposed upon the language of science.

Moritz Schlick took a different approach. Whereas Carnap had sought to begin with 'elementary propositions' and *construct* scientific interpretations from them, Schlick recast 'elementary propositions' as 'basic statements' whose role is to *terminate* the process of confirmation.[48]

Schlick's 'basic statements' have the form 'here, now, so and so'. They 'register' a confirming experience. For example, it may be hypothesized that an aqueous solution of the salt of a strong base and a weak acid will be basic. From this hypothesis, a hypothesis about the pH colour-dependence of litmus-paper, and statements about experimental operations, it may be deduced that a piece of red litmus-paper will turn blue. 'Here, now, blue' serves to mark the confirming experience.

'Basic statements' express the sense of a present gesture. According to Schlick, one understands the meaning of a 'basic statement' only as one executes the gesture. And to grasp the meaning of a 'basic statement' is to grasp its truth. 'Basic statements' contain the demonstrative terms 'here' and 'now'. Strictly speaking, 'basic statements' are not statements at all. They cannot be written down, since the inscription 'here' or 'now' destroys the immediacy of the gesture.

Hence 'basic statements' cannot serve as an evidential base for

---

[47] Carnap, 'The Methodological Character of Theoretical Concepts' in *Minnesota Studies in the Philosophy of Science* 1, ed. H. Feigl and M. Scriven (Minneapolis: University of Minnesota Press, 1956), 39.
[48] Moritz Schlick, 'The Foundation of Knowledge' in *Logical Positivism*, 209–27.

inductive generalization. They exist only in the activity of confirmation. Schlick declared that 'science does not rest upon them but leads to them, and they indicate that it has led correctly'.[49] He emphasized that the goal of science is success in this activity by which predictions are confirmed.

VII   MARGENAU'S 'CONSTRUCTIONIST' PHILOSOPHY OF SCIENCE

Henry Margenau combined a phenomenalist interpretation of Campbell's Hypothesis-Plus-Dictionary view of theories with a Kantian emphasis on regulative principles which prescribe the form of acceptable theories. The offspring of this marriage was a 'constructionist' philosophy of science.[50]

Margenau noted that certain elements of experience are characterized by spontaneity, passivity, irreducibility, and relative independence. These elements of experience include both sense impressions and dreams, hallucinations, and illusions. At the level of spontaneity and passivity, waking impressions and dreams are on a par. However, Margenau believed that a subdivision could be established within experience. Sense impressions, but not dreams and hallucinations, can be linked to constructs which satisfy certain formal and empirical requirements.

The first step in the transition from the data of immediate experience to scientific knowledge is the selection of 'rules of correspondence'. In Margenau's usage, 'rules of correspondence' link impressions to constructs. A variety of types of rules of correspondence are utilized in the formulation of scientific theories. One type associates sense data with an object which is posited as its source; 'external object' is a construct posited to account for regularities among our sense impressions. More abstract rules of correspondence correlate 'wavelength' with colour and 'electron path' with tracks observed in a cloud chamber.

The second step in the transition to scientific knowledge is the conversion of constructs into 'verifacts'. One condition that a

[49] Ibid., 223.
[50] Henry Margenau, *The Nature of Physical Reality* (New York: McGraw–Hill, 1950).

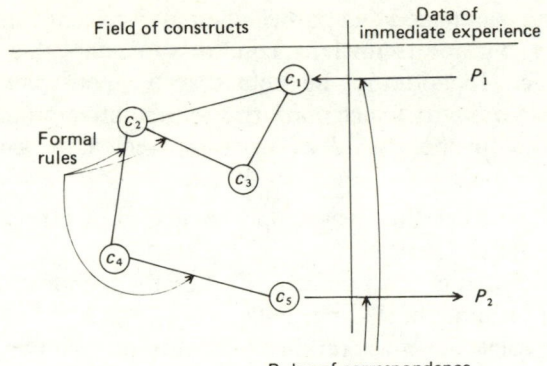

Fig. 4. Margenau's Circuit of Verification[a]

[a] Henry Margenau, *The Nature of Physical Reality*, 102–7.

construct must satisfy in order to qualify as a verifact is participation in circuits of verification (see Figure 4).

A circuit of verification begins and ends in the data of immediate experience. A transition from the perceptually given (P_1) to a construct (C_1) is accomplished by selection of a rule of correspondence. If C_1 has a use in science, it will be related formally to other concepts through axioms, theorems, and definitions. In virtue of these formal relations, it may be possible to recross the boundary between the abstract and the empirical by selecting a rule of correspondence that terminates at P_2. If P_2 corresponds to a datum of immediate experience, then the network of concepts traversed in the circuit has been confirmed. Margenau suggested that a necessary condition of the 'validity' of a theory, and the 'verifact-status' of its constructs, is that it is supported by a 'sufficient number' of circuits of verification.

Margenau thus reformulated the Hypothesis-Plus-Dictionary view of theories on a phenomenalistic basis. Dictionary entries link the terms of scientific theories to reports of the data of immediate experience.[51]

[51] The adequacy of a phenomenalistic reformulation was questioned by numerous philosophers of science. As Nagel pointed out, a phenomenalistic grounding of scientific theories is beset by two difficulties. In the first place, no precise linguistic framework is available for the use of phenomenalistic terms. And in the second place, the restriction of confirming instances to reports about 'colour images', 'tactile sensations', and the like would deprive scientific theories of their

Margenau maintained that participation in circuits of verification is a necessary, but not a sufficient, condition to establish a construct as a verifact. In addition, he held that the construct must be embedded in a theory which conforms to regulative principles that stipulate the 'proper form' of organized scientific knowledge, namely:

(1) The construct must be multiply connected by formal relations to other constructs;

(2) The construct must be related, directly or indirectly, to the data of immediate experience;

(3) The construct must retain uniformity of meaning in all its applications within the theory;

(4) The construct should be embedded in a theory from which laws of causal form may be derived;[52]

(5) The theory should be 'simple' and 'elegant'; and

(6) The concept should be susceptible to incorporation in additional theories.[53]

Margenau identified four constituents in scientific knowledge—the data of immediate experience, rules of correspondence, constructs, and regulative principles. Science progresses from data to constructs via selection of rules of correspondence. When this construction terminates in verifacts, the goal of science has been achieved. Moreover, reality has been created. Margenau proposed as a criterion of reality that, in all cases, if x is a verifact, then x is real. In so far as a concept functions in well-confirmed laws of causal form, it is a part of reality. From this point of view, the Ψ-function is every bit as 'real' as Mt. Everest. And the real world comprises

intersubjective testability (Ernest Nagel, *The Structure of Science* (New York: Harcourt, Brace & World, 1961), 120–9).

[52] Margenau held that a causal law is a law that specifies a determinate unfolding of the states of a physical system. From a causal law, together with information about the state of a system at one time, one can deduce information about the state of the system at future, or past, times. The mathematical form of a causal law is a differential equation which does not contain the time variable in explicit form. Such an equation is invariant with respect to the substitution of $(t - t_0)$ for t. Galileo's law of freely falling bodies is a law of causal form. From this law, together with information about the position and velocity of a body at one instant of time, one can predict the subsequent positions and velocities of the body.

Margenau recommended a 'Principle of Causality' as a regulative principle for the construction of scientific theories. This principle requires that physical systems be described by theories which imply laws of causal form.

[53] Margenau, *The Nature of Physical Reality*, 81–100.

all verifacts and those data of immediate experience which are correlated with verifacts by rules of correpondence.[54]

Margenau's version of the Hypothesis-Plus-Dictionary view of theories represented a marked shift of emphasis. Carnap, Frank, and Hempel had sought to anchor scientific theories in the bedrock of a physicalistic language. Their common assumption was that the truth or falsity of statements that assign values to the 'directly observable' properties of bodies may be ascertained independently of the truth or falsity of theoretical claims. They believed that the existence of a theory-independent observation language was necessary both to confer empirical significance on theoretical terms and to test theories in which these terms occur. From this standpoint, the important question about theoretical terms is how statements about such terms can be related to observation reports.

Margenau, on the other hand, did not accept this position on the existence of a theory-independent observation language. Instead, he gave a Kantian turn to the question about theoretical terms. For Margenau the important question was, 'given the possibility that some of our constructs are verifacts, what are the necessary and sufficient conditions of this possibility? The answer which he developed placed emphasis on empirical confirmation and regulative principles that stipulate the proper organization of empirical knowledge.

VIII POPPER

Members of the Vienna Circle had sought to formulate a criterion that would circumscribe the range of empirically meaningful propositions. Karl Popper judged this effort to be futile. He maintained that whether a proposition is empirically significant depends not on some alleged intrinsic 'meaning' that it possesses but rather on the way in which it is defended. The demarcation appropriate to science is thus between empirical methodology and various non-empirical approaches.[55]

Popper called attention to the problem of specifying criteria of demarcation for scientific interpretations. He noted that it is always possible to shield a pet theory from potentially falsifying

---

[54] Ibid., 299.
[55] Karl Popper, *The Logic of Scientific Discovery* (New York: Basic Books, 1959), 34–42, 81–7.

evidence. The position of conventionalism is unassailable on logical grounds.[56] Various strategies are available to restore agreement between a theory and evidence which appears to count against it. The simplest strategy is to reject the evidence. But even if the evidence is accepted, the theory may be qualified or modified so as to accommodate the evidence.

For instance, consider the case of a subject who has exhibited 'extra-sensory perception' in card-predicting experiments on Tuesday but fails to achieve predictive success on Wednesday. Does this falsify the hypothesis that the subject possesses extra-sensory perception? Not necessarily. One can save the ESP hypothesis by 'explaining away' Wednesday's results. Perhaps the subject had a headache. Or perhaps he sensed antagonism on the part of the person who turned over the cards. Or perhaps Wednesday's predictions were affected by a 'forward displacement' such that, unknown to the subject, these predictions refer not to the next card, but to the next-plus-one, or the next-plus-two, and so on.

Suppose someone offers as a 'scientific hypothesis' the claim that Jones has an extra-sensory predictive faculty, and then seeks to evade negative evidence by means of a wide range of strategies of the above type. At some point one becomes suspicious. The confrontations of hypothesis and evidence seem not to be genuine. Is there any evidence which could be specified that would be acknowledged to count against the hypothesis?

This is not to suggest that on every occasion of conflict between hypothesis and observational evidence the hypothesis should be rejected. Disputes may arise over the fruitfulness of particular strategies adopted to protect threatened hypotheses. What one scientist considers to be a refinement, or a natural extension, of a hypothesis, a second scientist may consider to be a misguided attempt to salvage a discredited hypothesis. An interesting case is the interpretation of the β-decay of radioactive nuclei. In β-decay, a nucleus expels a high-speed electron. Measurements revealed that the energy of the parent nucleus before β-emission is greater than the sum of the energies of the daughter nucleus, the electron, and any γ-rays emitted in the process. Scientists were loath to conclude that energy is not conserved in such processes. Wolfgang Pauli suggested in 1930 that a yet-undiscovered particle—the

[56] Ibid., 78–82.

neutrino—was emitted along with the electron. The neutrino was presumed to carry off just that amount of energy required to ensure conservation. It was not until 1953 that there was direct experimental evidence of the existence of this particle. Was the neutrino hypothesis in 1930 any less *ad hoc* than is the hypothesis of 'forward displacement of the predictions of ESP-subjects'?

Popper's position is that it is only with respect to the way in which hypotheses are defended that a satisfactory answer can be given to such a question. He recommended falsifiability as a criterion of demarcation for scientific interpretations. A hypothesis is scientific if, and only if, it is both logically possible and physically possible to falsify it. Present technical inability to design suitable tests need not disqualify a hypothesis. Tests of the neutrino hypothesis were physically possible, but not technically possible in 1930. Hence one could defend the neutrino hypothesis at that time while consistently seeking to expose the hypothesis to the strongest available tests.

For a hypothesis to qualify as scientific there must be some prospect for a confrontation with observational evidence. Nevertheless, the proponent of a hypothesis may respond to a negative test result by making changes elsewhere in his assumptions. In some cases he may be justified in making such a move. What he may not do is to respond consistently in this way. Popper conceded that one cannot specify a priori the point at which the defence of a hypothesis ceases to be scientific. However, he recommended that those concerned to promote empirical science undertake a commitment not to adopt evasive tactics on behalf of a pet theory. According to Popper the way to thwart conventionalism is to make a decision not to employ its methods.[57]

Hypotheses that satisfy the falsifiability criterion fall within the range of permissible scientific discourse. It remains to assess their acceptability. Popper's view is that acceptable hypotheses have proved their mettle by withstanding tests designed to discredit them. Tests of hypotheses are like the destruction tests which engineers perform to establish the strengths of materials.

Popper held that a test involves a confrontation between a hypothesis and a 'basic statement'.[58] Basic statements are singular statements that specify the occurrence of an intersubjectively

[57] Ibid., 82.   [58] Ibid., 47–8, 84–5.

observable event in a certain region of space and time. To test a hypothesis one must accept as true a basic statement describing a test result. But basic statements are not incorrigible. That a particular event occurred in a specific spatio-temporal region can be subjected to further tests. However, the testing of basic statements by reference to further statements must terminate at some point if the hypothesis itself is to be tested.

Popper acknowledged that, in the deductive testing of a hypothesis, we *decide* to accept certain basic statements for the purpose of the test. There is an element of conventionalism in the deductive method of testing. Popper declared that

the empirical basis of objective science has thus nothing 'absolute' about it. Science does not rest upon rock-bottom. The bold structure of its theories rises, as it were, above a swamp. It is like a building erected on piles. The piles are driven down from above into the swamp, but not down to any natural or 'given' base; and when we cease our attempts to drive our piles into a deeper layer, it is not because we have reached firm ground. We simply stop when we are satisfied that they are firm enough to carry the structure, at least for the time being.[59]

Popper introduced the phrase 'degree of corroboration' as an index of the fitness for survival of a scientific theory. Its magnitude increases with the increasing severity and diversity of tests successfully passed.[60] The severity of a test is difficult to measure. It depends on such factors as the ingenuity of the experimental arrangement, the accuracy and precision of the results achieved, and the relationship of the test result to rival hypotheses. In addition, much depends on how tightly the hypothesis tested is bound within a theoretical framework. If the hypothesis is so tightly bound that a negative test result could be accommodated only by a drastic revision of the system, then the test is a severe one. By this standard, the discovery that light rays from the stars are 'bent' by the sun, as predicted by Einstein's Theory of Relativity, is a test of high-level severity. By contrast, the discovery of yet another raven whose feathers are black is a test of low-level severity. In addition, a highly corroborated theory has passed tests in widely differing fields of application. For instance, the Theory of Relativity has passed tests that include the shift of spectral lines in vibrating atoms, the bending of light rays which

[59] Ibid., 111.   [60] Ibid., 265–9.

pass close to the sun, and an interpretation of a discrepancy in the orbit of Mercury.

Popper maintained that his position on corroboration reflects the growth-by-incorporation that often characterizes scientific progress. He noted that one theory is often superseded by a second theory which contains it, or an approximate version of it, and which has additional testable consequences.[61] This is the relationship between Newton's Mechanics and Einstein's Special Relativity Theory, for example. Presumably Popper takes the historical development of science to provide at least a partial justification for accepting his recommendation that 'degree of corroboration' be taken to be a criterion of acceptability for scientific theories.

[61] Ibid., 276.

5

Prescriptive Philosophy of Science and Inviolable Principles

I CHANGING EVALUATIVE STANDARDS AND THE CONTINUITY OF A PHILOSOPHY OF SCIENCE

OF course, it is not a decisive argument to claim that because philosophers of science have accorded prescriptive significance to the discipline, *PS* ought have prescriptive status today. Perhaps a persuasive case can be made for abandonment of the tradition of prescriptive *PS*. But tradition there has been. Eminent philosophers of science have accorded prescriptive significance to the discipline. The burden of proof rests on those who would promote a non-prescriptive alternative.

The prescriptive version of *PS* will be accepted provisionally at this point in order to investigate the necessary conditions of prescriptive status. The following two conditions appear to be necessary if philosophy of science P is to have prescriptive significance:

(1) P contains at least one evaluative standard S, and
(2) At least one application of S to a present evaluative situation in science determines *at that time* the correctness of a particular evaluative judgement in that situation.

A 'present evaluative situation' is a situation in which hypotheses, theories, or explanations—of whatever vintage—are assessed on the basis of whatever evidence is available at the present time. Condition (2) excludes from the class of prescriptive philosophies of science interpretations restricted to surveys of past evaluative practice.

The scope of application of S, *qua* evaluative *standard*, is more extensive than the set of evaluative situations to which it has been applied. Since S has an open scope, and since an application of S is judged to be correct, it seems appropriate to require that P contain a corresponding directive principle:

(3) Apply S to evaluative situations of the type cited in (2).

Of course, disputes may arise among practitioners of P about the correctness of classifying a particular situation as a situation of the type cited in (2).

The set of conditions—(1), (2), (3)—is too inclusive. It qualifies as a 'prescriptive' PS an interpretation which dismisses as irrelevant all demands for a justification of its evaluative standards. Such an interpretation qualifies as 'prescriptive' provided only that evaluative judgements based on present evidence are rendered and the corresponding directive principles are affirmed.

However, a 'PS' in which evaluative standards and their applications are not subject to justification is a PS quite different from those formulated by Aristotle, Whewell, Mill, Hempel, and Nagel. Each of these philosophers of science conceded the relevance of demands for the justification of evaluative standards and advanced justificatory arguments on behalf of the standards which they recommended.

If the set of necessary conditions for prescriptive status is to reflect the understanding of its eminent practitioners, then a fourth necessary condition is required:

(4) Questions about the adequacy of evaluative standards and particular applications of these standards are held to be relevant, and there exists at least one S, which satisfies conditions (2) and (3), for which a justificatory argument is given.

Every actual justification must terminate at some point—some 'non-justified justifiers' must be accepted. To justify an evaluative standard is to apply a higher-level evaluative principle, and to justify the higher-level principle is to apply a still higher-level principle. Since an infinite number of justificatory steps cannot be executed, there must be a highest level within any justificatory hierarchy that satisfies condition (4). At least one principle at this level will be unjustified.

Suppose, on the contrary, that every principle at this highest level is subject to justification. Given principles P_1, P_2, and P_3 at this level, one might seek to justify P_1, for instance, by application of P_2 or P_3. However, to do so would be to invoke the principle that 'an evaluative principle at the highest level of the justificatory hierarchy may be justified by application of other evaluative principles at the same (or lower) level'. This latter principle is a principle formulated at a higher level than that level which contains P_1. But, *ex hypothesi*, there is no higher level to which to appeal. Hence, there must exist within a prescriptive *PS* one or more evaluative principles not subject to justification.

The necessary conditions delineated thus far are conditions applicable to P at a given time. But P is an entity that persists over time and the question of its identity-over-time must be addressed.

It is a matter of historical fact that evaluative standards are often modified over a period of time. Of course one may elect to count any change in an evaluative standard as marking the origin of a new *PS*. However, to pursue this strategy is to complicate unnecessarily the history of the philosophy of science. A preferable strategy is to allow a certain type of change within P. Given evaluative standards S_1 at t_1 and S_2 at t_2, a plausible stipulation is that

(5) S_1 and S_2 are standards within one and the same P if, and only if, the transition from S_1 to S_2 satisfies the standards of rationality affirmed within P.

Standards of rationality themselves change over time. There are a number of ways in which condition (5) may be implemented. A strong version is:

(5*a*) S_1 and S_2 are standards within one and the same P if, and only if, the transition from S_1 to S_2 satisfies the standards of rationality affirmed within P at both t_1 and t_2.

However, if standards of rationality within P have changed during the time interval in question, a less stringent version may be appropriate:

(5*b*) S_1 and S_2 are standards within one and the same P if, and only if, the transition from S_1 to S_2 satisfies the standards of rationality affirmed within P at t_2.

Given that status as a prescriptive *PS* is contingent upon satisfaction of conditions (1), (2), (3), (4), and (5*b*), could there be a prescriptive *PS* in which no principles are inviolable? An inviolable principle is a principle whose rejection is tantamount to rejection of the *PS* in which it is a principle. A principle gains status as an inviolable principle only within a particular *PS*. The principle may subsequently be abandoned. But if it is abandoned, so too is the *PS* within which it had been affirmed to be inviolable. In this respect an inviolable principle has the same status within a normative *PS* as does a 'hard-core' principle in a Lakatosian scientific research programme.[1]

II SHAPERE'S PROGRAMME FOR A NON-PRESUPPOSITIONIST PHILOSOPHY OF SCIENCE

Dudley Shapere has recently sought to promote interest in a 'non-presuppositionist' *PS* that is normative, prescriptive, and devoid of inviolable principles. In studies undertaken in the 1970s, he sought to uncover methodological principles implicit in the development of scientific domains.[2] He concluded that a Principle of Discrete Compositional Reasoning was implicit in the development of theories about the periodically-ordered chemical elements and that a Principle of Evolutionary Reasoning was implicit in the development of theories about the spectral patterns of stars.[3]

Shapere recommended the above principles as principles directive of further research. If a domain satisfies specified conditions, then scientists *ought* to formulate theories which conform to the appropriate Principle of Reasoning. This, of course, is to move from a generalization about scientific practice to a conclusion about what counts as 'good science'.

Shapere failed to justify this transition in his 1970s essays. Subsequently he has admitted that in these essays he had recommended patterns of reasoning 'without any indication of their source or justification', and that a justification is needed needed.[4] Whatever the ambiguities of his earlier theory about domains, it is clear that Shapere's present project is to develop a

---

[1] See pp. 92–5 below, and also Imre Lakatos's essay 'Falsification and the Methodology of Scientific Research Programmes' in *Criticism and the Growth of Knowledge*, ed. I. Lakatos and A. Musgrave (Cambridge: Cambridge University Press, 1970), 132–8.

[2] Dudley Shapere, 'Discovery, Rationality and Progress in Science: A Perspective in the Philosophy of Science' in *Boston Studies in the Philosophy of Science* 20, ed. K. Schaffner and R. S. Cohen (Dordrecht: D. Reidel, 1974); 'Scientific Theories and Their Domains' in *The Structure of Scientific Theories*, ed. F. Suppe (Urbana: University of Illinois Press, 1974); 'The Influence of Knowledge on the Description of Facts' in *PSA 1976*, 2, ed. F. Suppe and P. Asquith (East Lansing: Philosophy of Science Association, 1977).

[3] See below, pp. 128–9.

[4] Shapere, *Reason and the Search for Knowledge* (Dordrecht: D. Reidel, 1983), xviii.

prescriptive *PS* in which evaluative principles are prescribed and justified.

In an essay on 'The Character of Scientific Change', Shapere suggested that 'we carry the historically-minded philosopher's insight to its fullest conclusion and maintain that there is absolutely nothing sacred and inviolable in science—that *everything* about it is in principle subject to alteration'.[5] Shapere declared that one goal of this *PS* is that the rationality of science be exhibited 'without presupposing criteria of what is to count as "rational", criteria which could not be *arrived at* in the course of seeking knowledge, but which must be assumed in order to engage in that enterprise at all, or at least to engage in it successfully'.[6]

There are three ways in which a *PS* might exhibit the rationality of science:

(1) by demonstrating that much of *HS* conforms to the rational reconstruction of scientific progress created by application of suprahistorical evaluative standards;

(2) by demonstrating that, although there are no suprahistorical evaluative standards, the transition from standard S_1 at t_1 to S_2 at t_2 is often rational; or

(3) by merely tracing the historical development of scientific practice.

Since Shapere denied that there exist suprahistorical evaluative standards, the first approach is unavailable. The third approach is also unavailable. It is possible that the rationality of science can be shown even if it cannot be demonstrated. But the showing is achieved, if it is achieved, by formulating a narrative faithful to scientific practice. And this task is a task for the historian of science. To select the third approach is to abandon *PS* and pursue *HS*.

Thus the second approach is the only viable approach by which Non-presuppositionist *PS* may exhibit the rationality of science. If Shapere's programme to develop a Non-presuppositionist, yet prescriptive, *PS* is to succeed, he must show that some transitions between evaluative criteria are justified.

Shapere maintained that 'there is often a chain of developments

---

[5] Shapere, 'The Character of Scientific Change' in *Scientific Discovery, Logic, and Rationality*, ed. T. Nickles (Dordrecht: D. Reidel, 1980), 63.

[6] Shapere, *Reason and the Search for Knowledge*, xxi.

connecting the two different sets of criteria, a chain through which a "rational evolution" can be traced between the two'.[7] If a transition from S_1 at t_1 to S_2 at t_2 is shown to be a 'rational evolution' of standards, then, *ceteris paribus*, it is rational to apply S_2, and not S_1, at t_2. And given a comparative (but provisional) justification of S_2, condition (4) of prescriptive status is satisfied.

A justificatory argument for the rationality of the transition from S_1 to S_2 invokes principle J below:

> J. S_2 is justified with respect to S_1
> if, and only if
> S_2 is superior to S_1 at t_2 by the standards of rationality of the PS at t_2.

But suppose S_1 is superior to S_2 at t_2 by the standards of rationality that were accepted at t_1. Consider, for example, the demarcation criterion observability-in-principle. Philosophies of science which emphasize that criterion are superior to those which do not by the standards of rationality of the Logical Empiricism of the 1930s. That judgement of superiority may not be consistent with standards of rationality of a present-day PS that reflects the emphases of quark theory.[8] But why not base comparative justification on the standards of rationality accepted at the earlier time?

One answer is that to do so is to do violence to our intuitions about the rationality of science. We associate the rationality of science and the historical development of science. To base justificatory procedures on standards of rationality that have been superceded is to uncouple the concepts 'rationality', 'temporal development', and 'progress'.

However, within a philosophy of science devoid of inviolable principles, a meta-level principle that stipulates that comparative justification of evaluative standards is to be determined by appeal to present-day standards of rationality is itself a provisional principle. Shapere has emphasized that what counts as a reason for accepting an evaluative standard is itself subject to change.[9] What are irrelevant considerations in one historical context may be bona

---

[7] Shapere, 'The Character of Scientific Change', 68.

[8] Ibid., 70–8.

[9] Shapere, *Reason and the Search for Knowledge*, xxix–xxx; 'Discussion of Shapere's Paper: "The Character of Scientific Change"' in *Scientific Discovery, Logic, and Rationality*, 102.

fide reasons in a different historical context. No a priori distinction between reasons and non-reasons is permitted.

Suppose a methodologist proposes a new meta-level criterion—that only those evaluative standards are acceptable whose applications are consistent with Immanuel Velikovsky's conclusions about the history of the solar system, for example that a comet ejected from Jupiter was responsible for a parting of the Red Sea and a temporary cessation of the Earth's rotation. Other methodologists may be expected to reject the newly proposed criterion on the grounds that meta-level criteria have not been selected previously by appeal to considerations of that sort. At its inception the Velikovsky supporter's recommendation violates standards of rationality currently accepted.

But perhaps this recommendation marks the origin of a significantly different context in which consistency with Velikovsky's conclusions is a newly relevant requirement. After all, if standards of rationality do change, there must be a time at which one standard displaces a second. The Velikovsky supporter's proposal cannot be dismissed out of hand.

Is there available within Non-presuppositionist *PS* a decision procedure to distinguish correct from incorrect present proposals about evaluative practice? It is clear that descriptive judgements may be rendered. For instance, it may be demonstrated that the Velikovsky supporter's proposal does not receive comparative justification on appeal to the standards of rationality accepted at the time the proposal is made. But to progress from that descriptive judgement to the claim that implementation of the Velikovsky supporter's proposal would be incorrect evaluative practice, and that this proposal ought not be adopted, is to invoke a higher-level directive principle such as D^*:

> D^* If a proposed revision of evaluative standards is inconsistent with the standards of rationality accepted within a *PS* at the time the proposal is made, then the proposal stipulates incorrect evaluative practice and ought not be implemented.

Given that standards of rationality can (and do) change, any particular application of D^* may be challenged. Defence of D^* requires appeal to a still higher-level directive principle such as D^{**}:

D^{**} Directive principle D^* is to be implemented until such time as its continued application is counter-productive.

How 'counter-productivity' is to be ascertained is not obvious. But, irrespective of that issue, there are additional difficulties.

In the first place, justificatory arguments about present evaluative practice involve a hierarchy of provisional directive principles, a hierarchy which involves principles like D^*, D^{**}, and its successors. Each directive principle is supported by reference to a principle at a higher level. But since no principle at any level is allowed to be inviolable within Non-presuppositionist PS, there is no apex to the hierarchy. A justification to establish the correctness of present evaluative practice cannot be completed. This is the case regardless of whether the proposal is for continuation or change of such practice.

Inability to produce a *complete* justification of evaluative standards does not disqualify Non-presuppositionist PS from the class of prescriptive philosophies of science. Condition (4) requires only that a justificatory argument be given for some evaluative standard, and a justificatory argument invoking, say, principle J, can be formulated within Non-presuppositionist PS.

But, even if a principle such as D^{**} could be justified, determination of the correctness of present evaluative practice would be contingent upon future developments. One cannot determine by consulting the principles of Non-presuppositionist PS at a given present time whether continuation or revision of current evaluative practice is correct. Non-presuppositionist PS does not satisfy condition (2), a condition taken to be necessary for prescriptive status.

Prospects for a Non-presuppositionist PS may improve if the strictures on prescriptive status are weakened. A proponent of Non-presuppositionist PS might object to condition (2) above. He might point out that appeal to provisional evaluative standards does introduce a distinction between correct and incorrect present evaluative practice, but that the considerations which determine correctness are future considerations. Suppose condition (2) is replaced by condition (2') below:

(2') At any given time, at least one application of S to a present evaluative situation in science permits a determination to be made *at some future time* about the correctness of a particular evaluative judgement in that situation.

A Non-presuppositionist *PS* may qualify as a prescriptive *PS* on this revised characterization. But it is a discipline in which promissory notes are issued without instructions for cashing them. Application of directive principle D**, for instance, requires that a decision be made about what counts as 'counter-productive' evaluative practice. Within Non-presuppositionist *PS*, what counts as 'counter-productivity' is subject to change with the growth of our experience.

But even if no disputes arise about the meaning of 'counter-productivity', there remains a vagueness about the assessment of evaluative practice. The methodologist wishes to know if it is correct to apply standard *S* in a particular present situation. Non-presuppositionist *PS* provides only 'pragmatic' counsel—'yes, if it works'. One can know whether 'it works' only by assessing future developments. But how extensive an assessment is one to make? No a priori stipulation about extensiveness is permitted within Non-presuppositionist *PS*. But then, a paraphrase of Feyerabend's query becomes pertinent—'if one is permitted to wait in reaching a decision about counterproductivity, why not wait a little longer?'[10]

Suppose, however, that a conclusion is actually reached that to have applied *S* yesterday would have been correct, given what has happened today. This conclusion, of course, was unavailable at the time of yesterday's deliberations. Nor does what has happened today establish that application of *S* would be correct evaluative practice today (the evaluative situation is no longer that of yesterday).

One can judge within Non-presuppositionist *PS* at time t_2 that evaluative practice at t_1 was correct, given the standards of rationality of some particular time, but one cannot justify at t_1 the claim that evaluative practice proposed at t_1 is correct. Since decisions about the correctness of evaluative practice are reached only after the fact, the 'normative' judgements of the philosopher of science are historical judgements. Indeed, the Non-presuppositionist philosopher of science is described fairly as a historian with a special interest in past evaluative practice in science.

If the above analysis is cogent, then the principle options are:

(1) to pursue the Non-presuppositionist programme under

---

[10] Paul Feyerabend, 'Consolations for the Specialist' in *Criticism and the Growth of Knowledge*, 215.

weak strictures on prescriptive status, and accept the consequent incorporation of philosophy of science into the history of science; or

(2) to stipulate within *PS* at least one principle held to be inviolable.

6

The Justificatory Hierarchy

THE inviolable standard required for a prescriptive *PS* may be introduced at various levels of the Justificatory Hierarchy (see Table 4).

Table 4. The Justificatory Hierarchy

Level	Content
1 Evaluative standards	Criteria of: demarcation confirmation acceptability completeness Schemata of causal relatedness Requirements for explanatory success *Logicism* Logical schemata Transhistorical criteria
2 Justification of evaluative standards	*Historicism* Case histories selected to exhibit applicability of evaluative standards Rational reconstructions of scientific progress compared to the history of science
3 Evaluative procedures for the justification of evaluative standards	Procedures, implementation of which are sufficient conditions of justification of evaluative standards
4 Justification of procedures for the justification of evaluative standards	Principles, satisfaction of which is a sufficient condition for the justification of evaluation procedures for the selection of a *PS*

I EVALUATIVE STANDARDS

Criteria of acceptability

At the lowest level of the Justificatory Hierarchy, philosophers of science may designate specific criteria to be inviolable. Among the criteria that have been proposed are agreement with observations, simplicity, conceptual integration, and fertility.

In retrospect, it may seem that these criteria have been ubiquitous within *HS*. But this is because the meanings of such terms as 'simple' and 'fertile' have changed over time. Moreover, what has been important to philosophers of science is not the individual criteria themselves but the way in which a balance is struck between them. For instance, a balance must be struck between conceptual integration and potential fertility. Conceptual integration achieved at the expense of potential fertility may be judged unsuccessful, particulary if *ad hoc* adjustments are used to secure agreement between theory and observation. A case in point is that version of phlogiston theory in which phlogiston, a substance supposedly exuded from metals during combustion, was assigned a 'negative weight'. This modified phlogiston theory accounts for the weight relations in combustion. However, the adjustment was purchased at too great a price—the uncoupling of the concepts 'substance' and 'weight'. The uncoupling of these concepts decreased the fertility of the phlogiston theory by de-emphasizing the importance of weight-determinations in the analysis of chemical reactions.

A balance seems necessary, as well, between agreement with observations and simplicity. If agreement with observations were all that mattered, it could be achieved by restricting scientific interpretations to collections of observation reports. But no philosopher of science would counsel scientists to restrict their enquiries to the collection and reporting of data.

The proposal to select a balance of conflicting criteria as an inviolable evaluative principle is initially an attractive proposal. Unfortunately the attractiveness diminishes at the point of specific determinations of the point of balance. Consider the determination of the appropriate balance between agreement with observations and simplicity in a case drawn from thermodynamics. Given three competing hypotheses—the Ideal Gas Law, the Van der Waals Law, and the Virial Expansion[1]—and data on the pressure–volume–temperature behaviour of gases, which hypothesis is the most acceptable? The Virial Expansion provides the best fit but is

---

[1] Ideal Gas Law: $PV = kT$

Van der Waals Equation: $\left(P + \dfrac{a}{V^2}\right)(V - b) = kT$

Virial Expansion: $PV = kT + \dfrac{A(T)}{V} + \dfrac{B(T)}{V^2} + \dfrac{C(T)}{V^3} + \ldots$

the most complex. The Ideal Gas Law is the most simple but provides only a loose fit. The Van der Waals Law provides a better fit than does the Ideal Gas Law and is less complex than the Virial Expansion. In actual practice, no one of the hypotheses is labelled 'most acceptable'. Scientists utilize whichever hypothesis suits the purpose at hand. But if the appropriate balance between agreement with observations and simplicity depends on the purpose at hand, then it is hard to see how this balance could be the inviolable principle required by prescriptive *PS*.

An inviolable principle within prescriptive *PS* warrants some (potential) present evaluative decisions. A balance of conflicting criteria qualifies as an inviolable principle only if a point of balance can be determined in specific evaluative situations and this point of balance in turn warrants a particular evaluative decision. A penumbra of vagueness is unavoidable in the determination of the appropriate point of balance. The suspicion arises that this penumbra is so wide that a wide range of evaluative decisions is consistent with the counsel to achieve a balance of conflicting criteria. If this is the case, then the balance principle is of limited value in actual practice.

The ideal of deductive explanation

Perhaps the appropriate inviolable standard is not some criterion of acceptability, or balance among criteria, but rather a standard of successful explanation. Aristotle, for instance, held that scientific explanation is a 'demonstration' in which a particular deductive form is instantiated. To explain an event or correlation is to display it in the conclusion of a 'properly formulated' deductive argument. 'Proper formulation' involves the satisfaction of both logical requirements—instantiation of syllogistic form *A A A* − I—and empirical requirements—that the premises be true, indemonstrable, better known than the conclusion, and state the cause of the attribution made in the conclusion.[2]

Aristotle's ideal of demonstrative knowledge was widely influential but did not carry the field. As early as the first century BC Geminus of Rhodes suggested that it might be fruitful to deduce statements about events or correlations from premises that include hypotheses

---

[2] Aristotle, *Posterior Analytics* 71^b 20–72^b 4.

not known to be true.[3] Andreas Osiander pursued this approach in his preface to Copernicus's *De revolutionibus*. Osiander suggested that the hypotheses of astronomers 'need not be true nor even probable; if they provide a calculus consistent with the observations, that alone is sufficient'.[4] Osiander noted that the position of Venus along the zodiac is deducible from premisses, one of which assigns an epicyclic motion to the planet. He emphasized that this premiss is clearly false since from it 'it necessarily follows that the diameter of the planet in the perigee should appear more than four times, and the body of the planet more than sixteen times, as great as in the apogee, a result contradicted by the experience of every age'.[5]

In 1948, Carl Hempel and Paul Oppenheim amended the Aristotelian ideal of explanation.[6] They maintained that one correct form of scientific explanation is suitable instantiation of the following deductive–nomological schema:

$$L_1, L_2, \ldots L_k \quad \text{General Laws}$$
$$\frac{C_1, C_2, \ldots C_r}{\therefore E} \quad \text{Statements of Antecedent Conditions}$$
$$\therefore E \quad \text{Description of Phenomenon.}$$

According to Hempel and Oppenheim, in a 'correct' deductive–nomological explanation, the premisses:

(1) are true;
(2) imply the conclusion;
(3) include general laws actually used in the deduction; and
(4) are subject to test by observation or experiment (at least in principle).

Eberle, Kaplan, and Montague[7] pointed out in 1961 that the Hempel and Oppenheim conditions for deductive explanation are too inclusive. Given a formal language which has the syntactic structure of the lower predicate calculus without identity, it is possible to construct within this language numerous arguments

---

[3] Geminus of Rhodes, quoted by Simplicius, in *A Source Book in Greek Science*, ed. M. Cohen and I. E. Drabkin (New York: McGraw–Hill, 1948), 91.

[4] Andreas Osiander, *Preface* to Copernicus, *De revolutionibus*, in *Three Copernican Treatises*, ed. E. Rosen (New York: Dover, 1939), 25.

[5] Ibid., 25.

[6] Carl G. Hempel and Paul Oppenheim, 'Studies in the Logic of Explanation', *Phil. Sci.* 15 (1948), 135–75; reprinted in Hempel, *Aspects of Scientific Explanation* (New York: Free Press, 1965), 245–95.

[7] Rolf Eberle, David Kaplan, and Richard Montague, 'Hempel and Oppenheim on Explanation', *Phil. Sci.* 28 (1961), 419–28.

that satisfy the Hempel and Oppenheim conditions but which ought not to count as 'explanations'. An example given by these authors is the following argument to explain why the Eiffel Tower conducts electricity:

> 'All mermaids conduct electricity.
> If it is false that the Eiffel Tower is a mermaid,
> then the Eiffel Tower conducts electricity.
>
> ∴ The Eiffel Tower conducts electricity.'

Given that no mermaids exist, the general law is true and the Hempel and Oppenheim conditions are satisfied. Since many accepted scientific laws are also vacuously true—for example the law of inertia, the law of the lever, and the Ideal Gas Law—it would seem that additional restrictive conditions are needed. Kaplan subsequently has developed a set of additional restrictions on instantiations of the deductive–nomological schema to block arguments of the Eiffel Tower type.[8]

The deductive ideal of explanation has clearly undergone changes from Aristotle to Kaplan. Is there, however, an essential aspect, or part, of this ideal that is a suitable inviolable Level 1 evaluative standard? For instance, would it be plausible to argue that deductive subsumption under laws is a sufficient condition of scientific explanation? Many contemporary philosophers of science would respond in the negative. Hempel himself has denied that deductive subsumption under laws invariably counts as scientific explanation. He called attention to the deductive–nomological arguments outlined below:

$$
\begin{array}{ll}
(1)\ \ T \propto \sqrt{l} & (2)\ \ T \propto \sqrt{l} \\
\quad\ l_2 = 1/4\, l_1 & \quad\ T_2 = 1/2\, T_1 \\
\therefore\ \ T_2 = 1/2\, T_1 & \therefore\ \ l_2 = 1/4\, l_1
\end{array}
$$

where T is the period of a pendulum and l is its length. We readily accept argument (1) as an explanation of the change of period of a pendulum. However, we are hesitant to accept argument (2) as an explanation of the change of length of the pendulum. Presumably this is because we can saw through a pendulum and thereby alter its period, but we cannot alter its period independently of changing its length. Hempel noted, however, that period and length are symmetrically related. It is true that we cannot alter its

---

[8] David Kaplan, 'Explanation Revisited', *Phil. Sci.* 28 (1961), 430–1.

period without changing its length. But it is also true that we cannot change its length without altering its period. Hempel declared that 'in cases such as this, the common-sense conception of explanation appears to provide no clear and reasonably defensible grounds on which to decide whether a given argument that deductively subsumes an occurrence under laws is to qualify as an explanation'.[9] Hempel also called attention to an argument formulated by S. Bromberger:

Laws	Theorems of physical geometry
Antecedent Conditions	Flagpole F stands vertically on level ground and subtends an angle of 45 degrees when viewed from ground level at a distance of 80 feet.

$\therefore$ Phenomenon   Flagpole F is 80 feet high.

Hempel noted that the premises of the flagpole argument do not explain why the pole happens to be 80 feet in length.[10] Similar considerations hold for arguments in which the occurrence of a storm is deduced from an 'indicator-law' that correlates storms with changes in readings of a barometer. Since an argument can fulfil the requirements of the Deductive–Nomological pattern without explaining its conclusion, Deductive–Nomological subsumption is not a sufficient condition of scientific explanation.

Of course, disqualification of any number of candidates for status as inviolable evaluative standards is consistent with the possibility that there do exist such standards. But until such time as a philosopher of science establishes a convincing case for a particular inviolable standard, the appropriate conclusion is that it is inappropriate to locate the permanence required for a prescriptive *PS* at this lowest level of the Justificatory Hierarchy.

II JUSTIFICATION OF EVALUATIVE STANDARDS

A more promising approach to the creation of a prescriptive *PS* is to locate the required inviolable criterion at Level 2 of the Justificatory Hierarchy. To do so is to claim that, although Level 1 evaluative standards are subject to modification and replacement

[9] Hempel, 'Deductive–Nomological vs. Statistical Explanation' in *Minnesota Studies in the Philosophy of Science* 3, ed. H. Feigl and G. Maxwell (Minneapolis: University of Minnesota Press, 1962), 109.
[10] Ibid., 109–10.

within a given *PS*, there is an inviolable Level 2 criterion by which Level 1 standards are to be evaluated.

Level 2 justificatory arguments involve an appeal to *HS*, to logical considerations, or both. The extreme standpoints are Unqualified Historicism and Unqualified Logicism. The Unqualified Historicist position is that historical considerations provide the sole warrant for evaluative standards. The Unqualified Logicist position is that logical, or perhaps 'broadly philosophical', considerations provide the sole warrant for evaluative standards.

It well may be that no philosopher of science ever defended an Unqualified Historicism or an Unqualified Logicism. However, William Whewell and John Stuart Mill came close. It was Whewell and Mill who first focused attention on the relationship between *PS* and *HS*. Their views on the ways in which *HS* is involved in the practice of *PS*, and vice versa, merit consideration.

Whewell and the Historicist standpoint

Whewell announced at the beginning of his *Philosophy of the Inductive Sciences* (first edition, 1840)[11] that he would base his *PS* upon an examination of *HS*. He proposed to survey the historical development of science and to draw conclusions therefrom about proper scientific method. He was well equipped to do this, since he had already published a three-volume *History of the Inductive Sciences* (first edition, 1837).[12] Whewell claimed originality for his approach, pointing out that previous writers on *PS* had cited historical episodes merely to support antecedently formed convictions about scientific method.

Whewell developed a prescriptive *PS*. He surveyed *HS* to find the evaluative criteria that are implied in scientific practice and then put forward these criteria as normative standards.

The principal criterion that he claimed to be exemplified in *HS* is the 'consilience of inductions'. A consilience of inductions is a conceptual integration in which less inclusive generalizations are incorporated into a more inclusive theory. Consilience is not achieved by a mere conjoining of laws. Rather, a superinduction of concepts is required. It is only from the perspective afforded by

---

[11] William Whewell, *Philosophy of the Inductive Sciences* (hereinafter *PI*) (2nd edn., 1847; London: Cass, 2 vols., 1967).

[12] Whewell, *History of the Inductive Sciences* (hereinafter *HI*), 3rd edn., 1857 London: Cass, 3 vols., 1967).

the concepts of a new theory that less inclusive generalizations are seen to be intimately related. Whewell declared that when consilience is achieved 'particulars form the general truth, not by being merely enumerated and added together, but by being seen *in a new light*'.[13]

Whewell noted that successive theoretical developments that achieve consilience display an increase in simplicity and coherence, and that developments that do not achieve consilience display an increase in complexity and disorder. He cited the development of the Corpuscular Theory of Light as a case of increasing complexity and disorder. Newton's theory had been successful in accounting for reflection and refraction. But to account for the colours of thin plates Newton was forced to tack on an assumption about 'fits of easy transmission and reflection'.[14] Subsequently, proponents of the Corpuscular Theory accounted for diffraction by subjecting the emitted particles to complex laws of attraction and repulsion; for polarization by postulating that emitted particles possess asymmetrical 'sides'; and for double refraction by subjecting particles to different forces along different axes of a crystal. Whewell maintained that, because the Corpuscular Theory had been augmented in this cumbersome, disharmonious manner, it was bound to be false.

By contrast, he eulogized the competing Wave Theory of Light. According to Whewell, the Wave Theory had been augmented and adjusted in a simple and harmonious manner. Both diffraction and the colours of thin plates were accounted for by making a single set of assumptions about the wavelengths of different colours of light, and polarization was accounted for on the assumption that the propagation of light produces transverse vibrations.

Whewell emphasized that undesigned scope is a sufficient condition of consilience. He declared that 'the evidence in favour of our induction is of a much higher and more forcible character when it enables us to explain and determine cases of a kind different from those which were contemplated in the formation of our hypothesis.'[15]

Whewell claimed that there is no instance in *HS* in which a

---

[13] Whewell, *PI*, 2. 85.
[14] Isaac Newton, *Opticks* (New York: Dover, 1952), 278–88.
[15] Whewell, *PI*, 2. 65.

theory displayed undesigned scope and subsequently was shown to be false.[16] He thus appealed to *HS* to justify the selection of an important evaluative criterion of *PS*. Whewell's justification procedure involves a 'burden-of-proof' shift. Potential critics are invited to find within *HS* a bona fide achievement of undesigned scope in which the theory in question was later discredited. Presumably, if a critic were to document such a case, then Whewell would de-emphasize or withdraw undesigned scope as a criterion of acceptability of theories.

Undesigned scope is only a sufficient condition of consilience. Whewell noted that consilience is often achieved in the absence of any manifestation of undesigned scope. He appealed to *HS* once again to justify selection of consilience as a criterion of theory replacement.

Whewell claimed that *HS* reveals that the evolutionary development of a science resembles the confluence of tributaries to form a river. Progress in science is growth by incorporation, in which past results are subsumed and reinterpreted by subsequent theories. Consistent with this finding, Whewell emphasized that rejected theories often contain elements that contribute to future progress. For example, Newton's correlation of colour and angle of refraction survived the demise of his Corpuscular Theory and was subsumed and reinterpreted by the victorious Wave Theory.[17] And the Phlogiston Theory's classification of combustion, respiration, and acidification as processes of the same kind was subsumed and reinterpreted by the victorious Oxygen Theory.[18]

Whewell's argument to justify the criterion of consilience may be reconstructed as follows.

1. The history of a science resembles a confluence of tributaries to form rivers.
2. Such confluence is invariably progressive.
3. If developments within a science exemplify a progressive pattern, then any criterion whose application in that context warrants that pattern is justified.
4. Application of the criterion of consilience warrants the Tributary–River pattern.

∴ Application of the criterion of consilience is justified.

The first premiss is a claim about a pattern exhibited within *HS*.

---

[16] Ibid., 2. 67–8.   [17] Whewell, *HI*, 2. 284–8.   [18] Ibid., 3. 103.

Whewell's justification of consilience is successful only if that premiss is true. But since the truth of the first premiss can be established only by historical enquiry, it might seem that Whewell's position is that *PS* is dependent on *HS* but not vice versa. This would be an extreme Historicist position. However, Whewell's full position on the relation of *PS* and *HS* is more complex.

Whewell recognized that in order to write *HS* it is necessary to judge the significance of various scientific developments. Reconstruction of the past requires evaluation and synthesis. Whewell selected a set of interpretive categories for the reconstruction of *HS*. At the most basic level, he posited a polarity of fact and idea, deciding in advance to interpret scientific developments in terms of this fundamental polarity.[19]

Ideas express those relational aspects of experience that are necessary conditions of knowledge. Ideas prescribe to experience; they are not derived from experience. Whewell included among 'fundamental ideas' both general concepts like space, time, and cause, and specific concepts fundamental to particular sciences. Among these specific concepts are 'elective affinity', 'polarity', and 'vital forces'.

In the *Philosophy of the Inductive Sciences*, Whewell maintained that

(1) basic to each science is a number of fundamental ideas;
(2) these ideas may be apprehended distinctly by the mind;
(3) the meaning of a fundamental idea may be unpacked in a set of axioms;
(4) an axiom may be expressed as a formal principle together with a statement ascribing content to the formal principle;
(5) a formal principle is an a priori, necessary truth; and
(6) the associated statement about content is specified within experience in the unfolding of the appropriate scientific discipline.

For instance, the idea of cause may be unpacked in a set of three axioms, the formal principles of which are (1) 'nothing takes place without a cause'; (2) 'effects are proportional to their causes'; and (3) 'reaction is equal and opposite to action'.[20] Whewell held that these formal principles are apprehended by the mind in such a way that no appeal to experience is relevant,[21] and no proof is either

---

[19] Whewell, *PI*, 1. 16–51.   [20] Ibid., 1. 177–85.   [21] Ibid., 1. 166.

required or possible.[22] He emphasized that it remained for the history of mechanics, culminating in Newton's laws of motion, to specify the content of the axioms of cause. Newton recognized that bodies possess no intrinsic cause of acceleration. He specified the manner in which impressed forces are compounded. And he identified the appropriate empirical quantities which qualify as 'action' and 'reaction'.[23]

Thus there is a further layer of complexity in Whewell's position on the relationship of *PS* to *HS*. Certain scientific laws express the empirical content of axioms about fundamental ideas. But recognition of the form of the axioms does not depend on developments within history.

A residual question is whether, and to what extent, Whewell's procedure is circular. Whewell defended the following propositions:

(1) a *HS* is formulated only if a *PS* is applied, and

(2) a *PS* is justified only if that *PS* is exemplified in a *HS*.

These propositions jointly imply:

(3) a *PS* is justified only if that *PS* is exemplified in an application of a *PS*.

This argument is not viciously circular. Whewell did not claim that the *PS* 'applied' in (1) is identical with the *PS* 'justified' in (2).

Nevertheless, the selection of philosophical categories to organize historical data may stack the cards in favour of the 'discovery' of certain kinds of evaluative criteria. Whewell begins with the Kantian concept of a set of categories which are a priori necessary conditions of cognitive experience and the Aristotelian concept of a series of distinct sciences, each with its appropriate subject matter, basic predicates, and first principles. It is not surprising that the *HS* which he formulates exhibits a Tributary–River pattern in which the 'fundamental ideas' of the individual sciences are progressively explicated. Nor is it surprising that his examination of the *HS* thus created reveals that the consilience of inductions is an important evaluative criterion.

Whewell's a priori commitment also undermines his justificatory argument for the criterion of consilience. The *HS* to which he appeals is itself an interpretation on behalf of an explicitly formulated philosophical position. The justificatory argument is

---

[22] Ibid., 1. 178.   [23] Ibid., 1. 215–45.

no stronger than his philosophically biased interpretation of science. And Whewell's reconstruction of *HS* is suspect. His commitment to the thesis that each science develops as an uncovering of relations among its 'fundamental ideas' led him to make some short-sighted claims about mid-nineteenth-century science.

In the case of optics, Whewell maintained that its fundamental ideas include 'propagation in a medium' and 'polarization', and that the wave theory of Young and Fresnel stipulates the *essential* relations of these concepts.[24] He emphasized the successes achieved by this theory and ignored its failures.

In the case of biology, Whewell maintained that the fundamental ideas include 'final cause' (in Cuvier's sense) and 'vital forces', but not 'random variations' or 'selective pressure'. He rejected Darwin's theory in spite of the consilience of biogeographical and palaeontological data which it achieved. Whewell rejected Darwin's theory in part because it was based on inappropriate ideas, in part because the evidence in its favour was inconclusive, and in part because it undercut the Argument from Design.[25]

Mill and the Logicist standpoint

John Stuart Mill disputed Whewell's claim that *HS* provides unequivocal testimony in support of the criterion of consilience. Mill noted that Whewell himself had documented a countercase. Whewell had admitted that Descartes's successors had modified the Vortex Theory of planetary motions to account for most of the facts explained by the rival Newtonian theory. Thus the Vortex Theory had achieved consilience prior to its eventual demise.

Mill misinterpreted Whewell at this point. Whewell had claimed only that no theory which unexpectedly accounted for a class of facts *without adjustment for that purpose* has been discredited.[26] Many theories have been adjusted to accommodate additional facts. And some theories, thus adjusted, subsequently have been discarded. But since Whewell did not maintain that every modified theory that accounts for additional phenomena is correct, Mill's criticism was not on target.

---

[24] Whewell, *PI*, 1. 315–17; *HI*, 2. 321–73.
[25] See M. Ruse, 'William Whewell and the Argument from Design', *Monist* 60 (1977), 224–68.
[26] Whewell, *PI*, 2. 67–8.

However, Mill did not rest his case against Whewell on supposed historical countercases. Mill's basic complaint was that Whewell's appeal to *HS* to justify evaluative criteria is wrong *in principle*. Mill maintained that all one can learn from *HS* is that certain regularities have held in the past, and that certain sequences of theories have exhibited consilience. Whewell was wrong—necessary status cannot be earned within history.

According to Mill, scientific enquiry is a search for causal connections, correlations that are both invariable and unconditional.[27] History may provide evidence that a correlation has been invariable; it cannot provide proof that the correlation will continue to be invariable, much less that the correlation is unconditional. It is logic, and not *HS*, that is the source of standards of verification in science.

Mill maintained that a law may be verified by demonstrating that it stands in a specific explanatory relationship to an empirically confirmed correlation of phenomena.[28] Given putative laws, L_1, L_2, . . . L_n and correlation p, a verification, or 'complete induction' conforms to the following deductive pattern.[29]

(1) L_1 explains p and p is true.
(2) L_1, L_2, . . . L_n are the only possible p-explainers.
(3) Neither L_2 nor L_3 nor . . . L_n explains p.
(4) If [(1) and (2) and (3)] then L_1 is true.
∴ L_1 is true.

In this context, Mill took 'explanation' to mean 'deductive subsumption'. He declared that 'a law of uniformity in nature is said to be explained when another law or laws are pointed out, of which that law itself is but a case, and from which it could be deduced'.[30]

Mill claimed that complete induction is often achieved in science.[31] But he cited just one instance—Newton's work on gravitational attraction. Mill and Whewell agreed that the Law of Gravitational Attraction had been verified. But whereas Whewell

---

[27] John Stuart Mill, *A System of Logic*) 8th edn., 1872; London: Longman, 1970), 221–2.
[28] The following criticism is intended to show that Mill's position on verification involves ambiguity and fails to achieve its purpose. This criticism does not rely on twentieth-century analyses of nomic status, explanation, or theory-comparison, which analyses pose additional difficulties for Mill's position.
[29] Mill, *A System of Logic*, 323.   [30] Ibid., 305.   [31] Ibid., 323.

held that verification involves an appeal to *HS* (which exhibits a 'consilience of inductions' created by Newton's theory), Mill insisted that verification is achieved only upon satisfaction of certain requirements of logic.[32]

Mill noted that Newton had demonstrated that Kepler's Law of Areas is a deductive consequence of premises that include a statement that a $1/r^2$ central force emanates from the sun. The Law of Gravitational Attraction thus explains the Law of Areas. Newton had also proved that no exponent other than -2 is consistent with the Law of Areas, thereby completing the induction.[33]

One necessary condition of complete induction is that the class of p-explainers may be 'examined'. Of course, if this class is finite, each may be studied in turn. But if the class of p-explainers is infinite this is not possible. In such a case it may be possible to eliminate all but one candidate by application of the theory of limits. For instance, Kepler's Law of Areas may be deduced from premises that include the relation $F\alpha\ 1/r^{2+\alpha}$ in the limit as $\alpha \to 0$. Since increasing positive or negative values of σ are associated with increasing divergence from Keplerian ellipses and the Law of Areas, it is not necessary to study the consequences of a great number of values of σ.

It would seem that the conditions of complete induction can be fulfilled only if the p-statement explained is unrestrictedly general with respect to space and time. Otherwise, indefinitely many laws may be formulated that deductively subsume p for the indicated region of space or time, but which have divergent consequences outside the restricted region.

One would expect Mill to hold that to fulfil the requirements of complete induction is to verify the relevant law. Indeed Mill declared that

we want to be assured that the law we have hypothetically assumed is a true one; and its leading deductively to true results will afford this assurance, provided . . . no law except the very one which we have assumed can lead deductively to the same conclusions which that leads to.[34]

However, in his discussion of Newton's complete induction for the $1/r^2$ central force law, Mill maintained that Newton's argument

[32] Ibid., 323.   [33] Ibid., 323–4.   [34] Ibid., 323.

would not have verified its conclusion had he not been entitled to assume that there was *some* deflecting force operating on each planet, a deflecting force, moreover, whose strength diminishes with increasing distance.[35] But since it was known that the planets were prevented from rectilinear motion by *some* centripetal force that decreases with increasing distance, Newton's argument did succeed in achieving verification.

Mill's point seems to be that without the assumptions of inertial motion and the existence of some centripetal force it would not be possible to examine the (possibly infinite) class of p-explainers. Mill may well have been under the spell of Newton's emphasis on *verae causae* at this point. Mill maintained that, for a complete induction to prove causal dependence, 'the supposed cause should not only be a real phenomenon, something actually existing in nature, but should be already known to exercise, or at least be capable of exercising, an influence of some sort over the effect'.[36] Mill insisted that he had no wish to restrict hypothesized causes to causes antecedently known to produce effects similar to those to be explained.[37] This disclaimer aside, Mill is guilty of conflating issues relevant to the discovery of causal connections with issues relevant to verification. According to Mill's own theory, satisfaction of the deductive schema for complete induction counts as verification, independently of any assumptions made in the course of formulating the requisite premises.

Be that as it may, complete induction is itself an unattainable ideal for science. If p is a generalization from observed data, then it is correctly expressed as an approximation, the limitations of which are specified by an associated probable error. An approximation—or for that matter even a set of data points—may be reproduced by a power expansion to any required degree of accuracy. The accuracy of the reproduction is limited only by the number of terms permitted in the expansion. If no limits are placed on complexity, a given approximation may be deductively subsumed by indefinitely many 'laws' of this type. It would not be possible to rule out all but one such 'law'. On the other hand, if the correlation expressed in p is applicable only to non-realizable, idealized situations—for example ideal levers, ideal gases, point-masses, and so on—then the complete induction is a contrary-to-fact claim.

---

[35] Mill, *A System of Logic*, 324.   [36] Ibid., 325.   [37] Ibid., 325.

Mill's reconstruction of Newton's complete induction is a case of this kind. Newton showed that if the sun and a planet were point-masses, if the sun's mass was infinite, if the sun and the planet were the only bodies in the universe, and if a planet obeys Kepler's Law of Areas, then this law follows from a $1/r^2$ central force emanating from the sun, and from no other force emanating from the sun. But the conjuncts in the antecedent clause are all false. In particular, planets do not obey Kepler's Law of Areas. The fit between the law and the observed motions of the planets is only approximate. Mill failed to note this discrepancy. Moreover he failed to note that Laplace and others had applied a theory of perturbations—itself based on Newton's Law—to account for the merely approximate fit between the Law of Areas and the motions of the planets.

Mill's position on verification, discussed above, is based on a deductive–nomological model of explanation. Mill also maintained that verification conforms to the schema of the inductive Method of Difference, namely:

Instance	Circumstances	Phenomena
1	A B C	a b c
2	B C	b c

$\therefore$ A is a necessary condition of a.

Mill declared that for the motion of a planet

A is a $1/r^2$ central force emanating from the sun.

BC are *all* the characteristics of the planet apart from exposure to the $1/r^2$ central force.

a is the direct proportionality between the line taken by the planet to traverse a segment of its orbit and the area swept out by the planet–sun line.

bc are *all* the motions of the planet other than a.[38]

Mill's subdivision of phenomena is suspect. It is not possible to study experimentally the motions that a planet would possess in the absence of the attractive force exerted by the sun. Mill noted that Newton obtained the correlations exhibited in instance (2) of the Difference schema by prior deduction rather than experiment.

[38] Ibid., 324.

He held that Newton did so by proving that, given circumstances BC, no force law other than A could produce a. Mill went on to claim that 'it is immaterial what is the evidence from which we derive the assurance that ABC will produce abc, and BC only bc; it is enough that we have that assurance'.[39] Mill maintained that Newton's instantiation of the schema counts as verification of the connection asserted in the conclusion.

But the two instances represented in the schema have been idealized. Neither Newton nor anyone else can determine BC, interpreted as *all* circumstances other than A. The best we can do is to hypothesize that specific circumstances A, B, and C are the only *relevant* circumstances. Mill's critics have correctly emphasized that the adequacy of the Method of Difference as an instrument of discovery is contingent upon the correctness of hypotheses about relevant circumstances. The critics have also insisted that the Difference schema is a rule of proof only if two premises are added: (1), A, B, and C are the only circumstances relevant to the occurrence of a, and (2) if A is a necessary condition of a phenomenon of type a on one occasion in which A, B, and C are the only relevant circumstances, then A is a necessary condition of a phenomenon of type a on every occasion that A, B, and C are the only relevant circumstances present.[40]

Once again Mill has developed an analysis applicable only to a non-realizable case. He was wrong to claim that Newton's $1/r^2$ Central Force Law is verified upon application of the Method of Difference. Verification cannot be achieved by instantiating the Difference schema with a small number of (types of) circumstances and phenomena. Nor is it possible to specify *all* circumstances present at the occurrence of a phenomenon. Mill's ideal of verification by instantiation of the Difference schema cannot be realized within science.

Mill was evidently unaware that he had formulated two distinct concepts of verification.[41] Verification in one sense is complete

---

[39] Ibid., 324.

[40] See, for instance, W. S. Jevons, *Pure Logic and Other Minor Works* (London: Macmillan, 1890), 295.

[41] Actually, Mill suggested three concepts of verification: (1) 'complete induction'; (2) 'complete induction' with the proviso that a *vera causa* be specified; and (3) satisfaction of the Method of Difference. In addition, as Ducasse points out, Mill also speaks of 'verification' in cases in which observation reports confirm the deductive consequences of hypotheses (Curt J. Ducasse, 'John Stuart Mill's

induction, an ideal based on a deductive model of explanation. Verification in the second sense is satisfaction of the Method of Difference. That he failed to note this equivocal usage is perhaps due to the way in which he reconstructed Newton's derivation of Kepler's first two laws. Mill obtained the requisite instances for the Difference schema from considerations about possible explanations of the Keplerian relations, thereby conflating the two senses of 'verification'.[42]

Mill provided a justificatory argument for the Method of Difference. He claimed that addition of a premiss that asserts the Uniformity of Nature converts the Difference schema into a valid deductive argument. Given that the positive and negative instances cited in the schema differ only in circumstance A, the appropriate additional premiss is that 'in every case in which circumstances of types A, B and C alone are present, a phenomenon of type a occurs'.

That 'Nature is Uniform' is itself the conclusion of an inductive argument by simple enumeration. Our evidence for uniformity is individual regularities that have been observed. Mill conceded that arguments by simple enumeration are risky. For one thing, the sample upon which the generalization is based may not be a representative sample. He maintained, however, that the reliability of arguments by simple enumeration increases as the scope of the generalization increases. In the limiting case of the conclusion 'Nature is Uniform', the scope of the generalization is *every* sequence of events. And historical enquiry reveals that 'no exception to uniformity has been documented'. Mill declared that every seeming exception 'sufficiently open to our observation' has been accounted for, either by finding that an additional circumstance was present or by finding that a type of circumstance usually

---

System of Logic' in *Theories of Scientific Method*, ed. R. M. Blake, C. J. Ducasse, and E. H. Madden (Seattle: University of Washington Press, 1960), 231). But this is an atypical usage in Mill. Indeed, Mill's usual position is to attribute this latter usage to Whewell and to repudiate it.

[42] In addition to the Method of Difference, Mill formulated inductive methods of Agreement, Concomitant Variations, and Residues. He claimed that it was application of the Method of Agreement that led to the conclusion that electrified bodies induce a contrary electrified state in nearby bodies (*A System of Logic*, 270), and application of the Method of Residues that led to the discovery of lithium in mineral sulphates (ibid., 281).

present was absent.[43] He concluded that the uniformity principle is 'duly and satisfactorily proved' by simple enumeration.[44]

Although Mill defended a Logicist position on the origin of Level I evaluative standards, he conceded that the justification of these standards requires appeal to *HS*. It is historical evidence that provides inductive support for the uniformity premiss.

Mill's justificatory argument is not successful. No appeal to observed regularities can prove that things could not be otherwise. Hume had emphasized this in a passage that anticipates Mill's argument:

All inferences from experience suppose, as their foundation, that the future will resemble the past, and that similar powers will be conjoined with similar sensible qualities. If there be any suspicion that the course of nature may change, and that the past may be no rule for the future, all experience becomes useless, and can give rise to no inference or conclusion. It is impossible, therefore, that any arguments from experience can prove this resemblance of the past to the future; since all these arguments are founded on the supposition of that resemblance. Let the course of things be allowed hitherto ever so regular; that alone, without some new argument or inference, proves not that, for the future, it will continue so.[45]

Hume is correct in advance. Mill failed to justify his claim that instantiations of the Method of Difference establish causal connections, that is connections both invariable and unconditional.

Lakatos on the evaluation of historiographical research programmes

Whewell and Mill proposed justificatory arguments on behalf of individual Level I evaluative standards. More recently justificatory arguments have been advanced on behalf of entire methodologies, comprising sets of evaluative standards.

Among the evaluative standards of a given methodology are criteria to gauge the comparative acceptability of scientific theories. Prescriptive *PS* thus has an interesting application to *HS*. A philosopher of science may select a sequence of theories from *HS*—T_1, T_2, T_3, . . .—and apply the criteria of acceptability of

[43] Ibid., 374.   [44] Ibid., 373.

[45] David Hume, *An Enquiry Concerning Human Understanding* in *Hume's Enquiries*, ed. L. A. Selby-Bigge, rev. P. H. Nidditch (Oxford: Clarendon Press, 1975), 37–8.

his methodology to successive pairs in the sequence. If the result is that T_2 is more acceptable that T_1 and T_3 is more acceptable than $T_2 \ldots$, then the sequence counts as a case of scientific progress. In so far as different philosophies of science stipulate different criteria of acceptability, they generate different rational reconstructions of scientific progress.

Interpretations of what counts as progress within *HS* are subject to evaluation at Level 2 of the Justificatory Hierarchy. Imre Lakatos has observed that 'all methodologies function as historiographical (or meta-historical) theories (or research programmes) and can be criticized by criticizing the rational historical reconstructions to which they lead'.[46] The criticism invokes *modus tollens* arguments of the form:

$$N \supset R$$
$$\frac{\sim R}{\therefore \quad \sim N}$$

where N = 'prescriptive *PS n* is correct'; and R = 'rational reconstruction r is the proper interpretation of developments within *HS*'.

To argue that one rational reconstruction is superior to a second is to argue that it is the better interpretation of scientific developments. A presupposition of such justificatory arguments is that there is a 'history of science' against which rational reconstructions may be compared. This 'history of science', in turn, is an interpretation of records that include books, letters, journal articles, symposium abstracts, and the like.

Prima facie, there is an element of circularity in attempts to evaluate rational reconstructions of scientific progress. Appeal is made to a 'history of science', but this 'history of science' is itself an interpretation which reflects a methodological bias. There is no philosophically neutral *HS*. How then can an appeal to *HS* be a satisfactory approach to the evaluation of competing methodologies?

Lakatos was well aware of this difficulty. Nevertheless, he developed a Level 2 evaluative procedure in which appeal to *HS* is an essential step. His procedure for the evaluation of rational reconstructions will be examined following a discussion of his position on Level 1 evaluative standards.

---

[46] Imre Lakatos, 'History of Science and Its Rational Reconstructions' in *Boston Studies in the Philosophy of Science* 8, ed. R. Buck and R. Cohen (Dordrecht: D. Reidel, 1971), 109.

The methodology of scientific research programmes Lakatos maintained that Level 1 appraisals ought be directed not at individual theories, but rather at 'scientific research programmes'. He declared that a scientific research programme 'consists of methodological rules: some tell us what paths of research to avoid (negative heuristic) and others what paths to pursue (positive heuristic)'.[47]

A research programme becomes implemented in a sequence of theories, each of which is a modification of its predecessor. The sequence develops over time in response to the positive heuristic of the programme. The positive heuristic stipulates (or at least suggests) what types of theory-change are appropriate in response to anomalies.

A research programme has both a static component and a dynamic component. The static component is a central core of laws and assumptions that are not exposed to falsification. The dynamic component is a protective belt of auxiliary hypotheses that are formulated in the course of applying the central core to additional phenomena (see Figure 5).

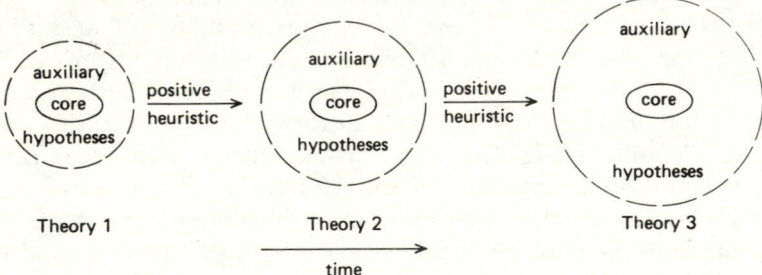

Fig. 5. Implementation of a Lakatosian Research Programme

To accept a research programme is to accept as inviolable a central core of laws and assumptions. Practitioners who do accept the research programme direct all tests at the protective belt of auxiliary hypotheses.

An example discussed by Lakatos is the following quasi-historical Newtonian Research Programme for astronomy.[48] Its

---

[47] Lakatos, 'Falsification and the Methodology of Scientific Research Programmes' in *Criticism and the Growth of Knowledge*, ed. I. Lakatos and A. Musgrave (Cambridge: Cambridge University Press, 1970), 132.

[48] Ibid., 135–6.

Table 5. Newtonian Research Programme

Theory	Auxiliary hypotheses	Results of applying theory
T_1	Sun stationary	Kepler's Laws deduced
	Sun and planet are point-masses such that $m_s \gg m_p$	Fit only approximate
T_2	Sun and planet move about common centre of gravity	Improved fit, but motions of Jupiter and Saturn are anomalous
T_3	Acknowledge perturbations	Fit further improved
	Seek approximate solutions to 3-body interaction	Anomalous motions of Jupiter and Saturn described by T_3
		Motion of Moon anomalous
T_4	Correction introduced for asymmetric mass-distributions	Motion of Moon described with improved accuracy by T_4
		Anomalous motion of Uranus noted as more data becomes available
T_5	A trans-Uranic planet exists	Neptune discovered near predicted location

central core includes the three axioms of motion and the law of gravitational attraction (see Table 5).

The commited Newtonian refused to abandon the central core of his programme upon discovery of anomalous motions. Instead, he added hypotheses to the protective belt. Lakatos emphasized that the augmentation of auxiliary hypotheses is often creative.

A research programme is creative provided that its theories fulfil the following conditions:

(1) T_n accounts for the previous successes of T_{n-1};
(2) T_n has greater empirical content than T_{n-1}; and
(3) Some of the excess content of T_n is corroborated.[49]

Lakatos labelled such research programmes 'progressive problem-shifts'. If a point is reached at which a successor theory does not fulfil the above conditions, then Lakatos labelled the problem-shift 'degenerating'.

Lakatos noted that a research programme that is 'degenerating' at a particular time may subsequently stage a comeback. His favourite illustration of a resurgence of vitality is Prout's research programme for chemistry (1815).[50] One aim of Prout's programme was to show that the atomic weights of the elements are exact multiples of the atomic weight of hydrogen (1.0 g/g atom). Initially

[49] Ibid., 116–18, 134.   [50] Ibid., 138–40.

the programme seemed promising. As improved techniques became available for the determination of atomic weights, many newly determined weights approached whole-number values. But is also became increasingly evident that certain atomic weights— notably chlorine (35.5)—were not integral multiples of that of hydrogen. Most chemists abandoned the Proutian programme at that point. After the turn of the century, however, it was discovered that many elements occur in nature as mixtures of isotopes. Naturally occurring chlorine, for instance, is a mixture of Cl^{35} and Cl^{37}. The Proutian programme was revived and applied to isotopes of elements rather than to elements themselves.

Feyerabend complained that application of Lakatos's criterion of progress may actually thwart progress in some cases.[51] If a research programme judged degenerate at a particular time is in fact the first stage of a long-term progressive programme, then to abandon the programme is to lose a chance to achieve progress. On the other hand, if the programme subsequently continues to degenerate, continued commitment to the programme would be counter-productive.

In reply, Lakatos insisted on a distinction between the act of appraisal and the decision whether or not to continue working within the scope of a research programme. The act of appraisal is an *objective* application of evaluative criteria, even though the appraisal verdict may be different at different times. Given a negative appraisal at a given time some scientists may decide to persevere in their support of a research programme. It is not necessarily irrational to pursue a currently non-progressive research programme. Lakatos declared that 'it is perfectly rational to play a risky game: what is irrational is to deceive oneself about the risk'.[52]

Feyerabend also complained that Lakatos's criterion of progressive problem-shifts has a very limited range of application. According to Lakatos, a research programme is progressive if each of the theories incorporates its predecessor and achieves some empirical support for its additional content. Feyerabend held that this is an idealization seldom found in *HS*. The usual relationship when one theory replaces another is that:

[51] Paul Feyerabend, 'Consolations for the Specialist' in *Criticism and the Growth of Knowledge*, 215–16.
[52] Lakatos, 'History of Science and Its Rational Reconstructions', 104 n.

(1) T_n accounts for some, but not all, successes of T_{n-1}; and
(2) T_n accounts for an additional range of facts not explained by T_{n-1}.[53]

For instance, although the Newtonian Research Programme initially explained Kepler's Laws in a way that the Cartesian Vortex Theory could not match, the Vortex Theory initially explained the unidirectional revolutions of the planets in a way that the Newtonian Programme could not match. Feyerabend emphasized that since Lakatos's criterion of comparative acceptability cannot be applied to cases of explanatory overlap, it is of little use as an evaluative standard for scientific theories.

A Kuhnian evaluative criterion In the course of a defence of Thomas Kuhn's *PS*, Jack Meiland suggested a Level 1 criterion of comparative acceptability that is applicable to cases of overlap.[54] Kuhn had maintained that a competition between paradigms[55] cannot be settled by appeal to paradigm-neutral observation reports. Observational evidence is viewed differently from the standpoints of different paradigms. Tycho sees the rise of the sun; Kepler sees rotation of the earth beneath the stationary sun. A Newtonian scientist sees the (nearly) isochrynous motion of a

[53] Feyerabend, *Against Method* (London: NLB, 1975), 176–80.
[54] Jack Meiland, 'Kuhn, Scheffler, and Objectivity in Science', *Phil. Sci.* 41 (1974), 179–87.
[55] Kuhn conceded in a 'Postscript' to the second edition of *The Structure of Scientific Revolutions* (1970) that he had used the term 'paradigm' equivocally in the First Edition (1962) (Thomas S. Kuhn, *The Structure of Scientific Revolutions* (2nd edn., Chicago: University of Chicago Press, 1970), 174–91).

In its narrow sense, a 'paradigm' is an 'exemplar', a 'concrete puzzle solution' which provides clues for the solution of current research problems. Examples include Newton's application of the theory of gravitational attraction to the motions of the planets, and Mendeleeff's prediction of the properties of hitherto unknown elements from his periodic classification system.

In its broad sense, a 'paradigm' is a 'disciplinary matrix', a shared commitment to all or several of the following:

(1) types of procedure—e.g. *in vivo* studies rather than *in vitro* studies;
(2) evaluative criteria—e.g. simplicity, agreement with observations, fertility;
(3) patterns of explanation—e.g. deductive–nomological, statistical–inductive, teleological;
(4) the existence of theoretical entities—e.g. atoms, fields, micro-organisms; and
(5) one or more 'paradigms' in the narrow sense.

It is competition between 'paradigms' *qua* 'disciplinary matrices' that is the subject of Meiland's analysis.

pendulum; an Aristotelian scientist sees a constrained body thwarted in its downward motion toward its 'natural place'. Kuhn suggested that adherents of competing paradigms practise science 'in different worlds'. He declared that 'the two groups of scientists see different things when they look from the same point in the same direction'.[56] Kuhn thus held that paradigm replacement is rather like a *Gestalt*-shift.

Critics of Kuhn, among them Lakatos, wrongly accused Kuhn of defending an irrationalist position on scientific revolutions in which one paradigm displaces a second. Meiland pointed out, however, that Kuhn *had* recommended objective criteria for the evaluation of competing paradigms. Kuhn had suggested that a successful new competitor should deal constructively with problems that the previous paradigm had solved. If a new paradigm does this and, in addition, solves problems not solved by the earlier paradigm, then the new paradigm is superior.

But there is a further problem. How do we know whether a problem formulated within a paradigm is 'solved' by the paradigm? The criteria of problem solution themselves may be paradigm-dependent. Meiland insisted that, even if what counts as a 'solution of a problem' is different for each paradigm, there yet may be a paradigm-independent criterion for the evaluation of paradigms, namely

P_2 is superior to P_1 if P_2 solves more of the problems it generates (according to the criteria of solution of P_2) than P_1 solves of its problems (according to the criteria of solution of P_1).[57]

This is an *objective* criterion of paradigm selection, a Level 1 standard applicable to competing paradigms independently of a possible relation of overlap between them. Unfortunately, application of the Kuhn–Meiland criterion may lead to counter-intuitive results. On the questionable assumption that problems may be individuated clearly within each competing paradigm, one paradigm may earn comparative justification, not on merit, but because its standards for what counts as a 'solved problem' are the more generous.

---

[56] Ibid., 150.
[57] Meiland, 'Kuhn, Scheffler, and Objectivity in Science', 183.

Evaluation of competing methodologies Lakatos applied the formal criterion 'incorporation with corroborated excess content' to assess both scientific research programmes and historiographical research programmes. In each case his focus is on sequences of interpretations. In the case of historiographical research programmes, one selects a sequence of methodologies and elaborates the rational reconstructions of scientific progress implied by each methodology. Next, one compares each rational reconstruction against 'the history of science'. If successive reconstructions are increasingly more congruent with 'the history of science', then the sequence is progressive. And given that a sequence is progressive, one ought to accept and apply the theory of scientific method that stands at the proximate end of the sequence. There is, then, an isomorphism of structure in the appraisal of scientific research programmes and historiographical research programmes (see Table 6).[58]

Lakatos's appraisal procedure for historiographical research programmes involves appeal to both *HS* and an inviolable suprahistorical evaluative principle. He applied this evaluative procedure to the sequence—Inductivism–Popperian Methodological Falsificationism–Lakatosian Methodology of Scientific Research Programmes. He maintained that the Popperian rational reconstruction of scientific progress is superior to Inductivism because it restores to scientific status certain discredited theories. For example, Popper's reconstruction successfully 'rehabilitated the scientific status of falsified theories like phlogiston theory, thus reversing a value judgment which had expelled the latter from the history of science proper into the history of irrational beliefs'.[59]

Lakatos maintained, in addition, that his own methodology is superior to Popper's falsificationism. One point of superiority is that the methodology of scientific research programmes accounts for the 'relative autonomy of theoretical science'.[60] Lakatos emphasized that, in the practice of theoretical science, research programmes are often pursued in the face of dramatic falsifications. An important example is the nineteenth-century defence and

---

[58] Lakatos, 'Falsification and the Methodology of Scientific Research Programmes', 116–18, 132–8.

[59] Lakatos, 'History of Science and Its Rational Reconstructions', 117.

[60] Lakatos, 'Falsification and the Methodology of Scientific Research Programmes', 137; 'History of Science and Its Rational Reconstructions', 99.

Table 6. Lakatos on the Appraisal of Scientific Research Programmes and Historiographical Research Programmes

	Scientific research programmes	Historiographical research programmes
Unit of appraisal	A sequence of scientific theories (e.g. development of the Newtonian programme)	A sequence of rational reconstructions (e.g. Inductivism–Popperian Methodological Falsificationism –Lakatosian Evaluation of Scientific Research Programmes)
Rules of appraisal	Given a sequence of scientific theories—T_1, T_2, . . . T_r—if (1) T_n accounts for the previous successes of T_{n-1}; (2) T_n has greater empirical content than T_{n-1}; and (3) Some of the excess content of T_n is corroborated; then the sequence is progressive.	Given a sequence of rational reconstructions—R_1, R_2, . . . R_r—if (1) R_n accounts for the successes of R_{n-1}; (2) R_n reconstructs some historical episodes that do not conform to the standards of rationality of R_{n-1}; and (3) Some of the excess content of R_n is corroborated; then the sequence is progressive.
Limitations of appraisal	Time factor A degenerating shift may subsequently stage a comeback An experiment is 'crucial' only in retrospect	Any rational reconstruction does less than full justice to the actual history of science A rational reconstruction must be supplemented by an 'external' historical account

elaboration of the Newtonian research programme in the face of anomalous data on the orbit of Mercury. On the Popperian reconstruction, such episodes are excluded from the rational growth of science. The Lakatosian reconstruction of scientific growth, by contrast, can accommodate episodes of this type.

A second point of superiority is that the methodology of scientific research programmes can accommodate theories formulated on inconsistent foundations. Lakatos cited Bohr's 1913 Theory of the Hydrogen Atom. Bohr used the Maxwell–Lorentz theory of electromagnetism to calculate the energy of the bound electron, but insisted, contrary to the Maxwell–Lorentz theory, that the electron revolves in an orbit without radiating energy.[61] According

---

[61] Lakatos, 'Falsification and the Methodology of Scientific Research Programmes', 140–54.

to Lakatos, Popperian methodological falsificationism cannot rationally reconstruct such episodes. A theory with inconsistent postulates implies everything, and so there can be no genuine test of its deductive consequences.[62]

There are historical episodes, then, that are reconstructed by the Methodology of Scientific Research Programmes but not by Methodological Falsificationism. Given that the Methodology of Scientific Research Programmes also accounts for all the historical episodes reconstructed by Methodological Falsificationism—a claim not demonstrated by Lakatos—the Methodology of Scientific Research Programmes qualifies as superior on Lakatos's criterion of incorporation with corroborated excess content.

Lakatos's evaluation procedure is open to a charge of circularity. Lakatos held that each *PS* generates a rational reconstruction of scientific progress by marking off those episodes that fit its ideal of rationality from those episodes that do not (the 'external' history of science). He also held that if one methodology renders rational all aspects of *HS* that are rational according to a second methodology, and more besides, then the first methodology is superior. But, as has been emphasized, every *HS* is an *interpretation* of the available evidence, an interpretation undertaken from a particular standpoint. Lakatos judged his own Methodology of Scientific Research Programmes to be superior to both Popperian Falsificationism and Inductivism on the basis of an appeal to a *HS* formulated according to the canons of the Methodology of Scientific Research Programmes. The appraisal process is biased in favour of the methodological commitments of the appraiser.

Lakatos conceded this point. He maintained that, in order to exhibit the rationality implicit in the history of science, it may be necessary either to rearrange events within a sequence or to stipulate what scientists *should have* done or thought at a particular point in time. What results on this treatment is an idealized restatement of historical developments, supplemented by a collection of footnotes that indicate what actually did happen.[63]

Noretta Koertge suggested that Lakatos sought a 'Galilean

---

[62] Lakatos, 'History of Science and Its Rational Reconstructions', 112–13.

[63] Lakatos, 'Falsification and the Methodology of Scientific Research Programmes', 138–54.

Reconstruction' of the history of science.[64] The reconstruction is 'Galilean' because it is a result of extrapolation and idealization. Just as Galileo formulated laws applicable to 'free fall in a vacuum' and 'ideal pendulums', so Lakatos sought to create a history of science 'purified' by the elimination or modification of complicating circumstances.

Thomas Kuhn complained that Lakatos distorted the historical record in the process of building a case for the superiority of the Methodology of Scientific Research Programmes.[65] According to Kuhn, Lakatos's normatively reformulated 'history of science' is an inappropriate standard against which to appraise competing methodologies.

Kuhn has a legitimate complaint. Lakatos's evaluative procedure does appraise competing methodologies upon appeal to a *HS* formulated to reflect the canons of one of the competitors. However, this is an 'open-ended' circularity. It is true that to appraise as 'progressive' a sequence of rational reconstructions—R_1, R_2, R_3, . . .—is to appeal to a 'history of science' reformulated to reflect the canons of R_3. However, this does not mean that R_3 is a definitive rational reconstruction. A philosopher of science may subsequently create a methodology that implies R_4 such that the sequence—R_1, R_2, R_3, R_4—qualifies as 'progressive' upon appeal to a 'history of science' formulated to reflect the canons of R_4.

Thus Lakatos's own Methodology of Scientific Research Programmes is subject to justified replacement. However, the criterion by which the best methodology at a given time is selected—incorporation with corroborated excess content—is not itself subject to abandonment or modification within the history of science. It is by application of this inviolable criterion at Level 2 of the Justificatory Hierarchy that Lakatos insures prescriptive status for his philosophy of science.

Lakatos's criterion is inapplicable to cases in which there is an overlap of rational reconstructions. If the rational reconstruction generated from one methodology fits episodes not reconstructed by a second methodology, but the rational reconstruction generated from the second methodology also fits episodes not reconstructed by the first methodology, then there is no basis upon which to

---

[64] Noretta Koertge, 'Rational Reconstructions' in *Boston Studies in the Philosophy of Science* 39. 362–7.

[65] Kuhn, 'Notes on Lakatos' in *Boston Studies in the Philosophy of Science* 8. 143.

judge one methodology to be superior to the other. There is reason to believe that a partial overlap of rational reconstructions is not unusual.

Consider the rivalry between the Methodology of Scientific Research Programmes and Inductivism. Lakatos claimed that his own methodology reconstructs episodes not reconstructed by Inductivism (for example the history of theories about phlogiston).[66] It would seem, however, that late sixteenth/early seventeenth-century research on the 'amber effect' fits the Inductivist reconstruction and not the reconstruction suggested by the Methodology of Scientific Research Programmes.[67] Scientists such as Jerome Cardan, William Gilbert, and Niccolo Cabeo sought to implement Roger Bacon's 'Second Prerogative'—the use of experimentation to augment the range of facts upon which inductive generalizations are to be based.[68] This is not to say that these investigators were devoid of theories about the nature of the 'amber effect'. They clearly did have theoretical commitments. But their extensive testing to discover which substances display electrostatic effects fits an Inductivist reconstruction rather well. By contrast, these investigators did not share a body of 'hard-core' assumptions and did not develop an evolving collection of 'protective auxiliary hypotheses' in the process of implementing a 'positive heuristic'.

Consider also the rivalry between the Methodology of Scientific Research Programmes and Popperian Falsificationism. Historical episodes in which an observation was accepted as decisive against a hypothesis are more readily accommodated to the Falsificationist methodology. Examples include Harvey's experiments against the hypothesis that there is an ebb and flow of blood in the circulatory systems of mammals, Hale's experiments against the hypothesis that there is a circulation of sap in plants, Count Rumford's experiments against the Caloric Theory of Heat, Davy's experiments against the oxygen theory of acids, the experiments of Foucault and Fizeau against the Newtonian Corpuscular Theory of Light, the Michelson–Morley experiment against the Ether Theory, and the Geiger–Marsden experiments against J. J. Thomson's 'Plum

---

[66] Lakatos, 'History of Science and Its Rational Reconstructions', 117.

[67] See, for instance, J. L. Heilbron, *Electricity in the 17th and 18th Centuries* (Berkeley: University of California Press, 1979).

[68] Roger Bacon, *The Opus Majus*, trans. R. B. Burke (New York: Russell and Russell, 1962), 2. 615–16.

Pudding' Model of the atom. Each of these episodes is subject to rational reconstruction on behalf of the categories of the Methodology of Scientific Research Programmes, particularly if one allows 'evolving' central cores and 'positive heuristics' that merely provide vague hints for theory development. However, it is arguable that a Methodological Falsificationist interpretation is the more convincing reconstruction.

Hence it is unclear that the Methodology of Scientific Research Programmes achieves comparative justification in competition with Inductivism and Popperian Methodological Falsificationism. It is likely that the relation between these methodologies is one of overlap rather than one of incorporation. If this is the case, then Lakatos's criterion is of limited use as an inviolable principle at Level 2.

Of course, one might augment the incorporation criterion by adding a clause to deal with cases of overlap, for instance:

methodology M_1 is superior to M_2, given a partial overlap between them,

if, and only if, the interpretations reconstructed by M_1 but not M_2 are more important than the interpretations reconstructed by M_2 but not M_1.

However, implementation of this clause would involve a serious element of circularity, since judgements of the relative importance of interpretations reconstructed are rendered from the standpoint of the methodology that is presumed to be superior.

But even if the comparative superiority of the Methodology of Scientific Research Programmes were incontrovertible on the criterion of incorporation with corroborated excess content alone, the Lakatosian Level 2 justification procedure is subject to an objection formulated by Laudan. Laudan noted that 'if we take his (Lakatos's) proposal seriously, then the best possible model of rationality would be that one which resulted in the judgment that every decision ever made in the history of science was rational'.[69] As Laudan pointed out, Lakatos's approach to justification is to take the entire history of science—reconstructed according to the canons of the latest, most inclusive methodology—as the standard of rationality. According to Laudan, this is a counter-intuitive

---

[69] Larry Laudan, *Progress and Its Problems* (Berkeley: University of California Press, 1977), 163.

standard. He declared that 'any model of rationality which made the *whole* of science rational would be as suspect as those models which make *none* of science rational'.[70]

Lakatos had conceded that no methodology, including his own, renders rational the whole of science.[71] But he took this to be an empirical claim. Laudan still is correct to insist that, *if* a methodology were formulated such that the whole of science, reconstructed according to its standards, qualified as rational, then it would be the winner of the justificatory sweepstakes. According to Laudan, this is counter-intuitive. But is it?

Laudan's criticism would have force only if there were some way to identify the 'whole of science' without reference to the standards of the latest, most inclusive methodology. But the demarcation of science from non-science at any given time is determined by appeal to just these standards. In so far as a methodology renders rational the 'whole of science', it does so, in part, by determining what counts as 'science'.[72]

As noted previously, the Lakatosian Level 2 evaluation procedure is open-ended. It always remains possible that a new methodology be developed which qualifies as rational both that 'whole of science' rendered rational by its predecessor, and additional developments as well. Of course, a reinterpretation of what counts as the 'whole of science' may be required.

Prospects for a Moderate Historicism

Lakatos's position on Level 2 evaluation reveals a strong commitment to the Logicist viewpoint. The criterion 'incorporation with corroborated excess content' is a formal criterion of acceptability.

The position of Unqualified Historicism, by contrast, is that proposed rational reconstructions are justified solely by appeal to *HS*. The practitioner of Unqualified Historicism would assess the comparative merits of the Inductivist and the Hypothetico-Deductive models of scientific progress, for instance, by super-imposing each model upon *HS*. The model which achieves the better fit would gain comparative justification. The Achilles' heel

---

[70] Ibid., 163.

[71] Lakatos, 'History of Science and Its Rational Reconstructions', 102.

[72] Laudan avoids this difficulty in his own appraisal procedure by focusing not upon what is believed to be the whole of science but upon selected important episodes which any reasonable demarcation would include within science.

of this Unqualified Historicism is the unavailability of a methodologically neutral *HS* upon which to project the rational reconstructions implied by competing philosophies of science.

A number of philosophers of science have sought to formulate a 'Moderate Historicism' intermediate between the unsatisfactory extremes of Unqualified Historicism and Unqualified Logicism. G. H. Merrill has characterized Moderate Historicism as a position in which 'neither purely historical nor purely *a priori* considerations alone determine the acceptability of a philosophical analysis of science'.[73]

James R. Brown has developed a Moderate Historicist position that is based on a two-part criterion of acceptability.[74] The first part of the criterion involves an appeal to *HS*. It is couched in terms of a distinction between 'theoretical' and 'normative' reconstructions of historical episodes. A theoretical reconstruction is a 'description' of historical episodes in the categories of some methodology, for example 'research programme', 'hard core', 'auxiliary hypothesis' (Lakatos). Brown emphasized that it is not possible to formulate a *HS* without employing the conceptual apparatus of some *PS* or other. However, the intention of a theoretical reconstruction is to recapture the actual course of development of science. A normative reconstruction, by contrast, stipulates how a historical episode *ought* to have developed according to some methodology. Given methodologies *M* and *m* which yield theoretical reconstructions of historical episodes *T* and *t* and normative reconstructions of historical episodes *N* and *n*, the first part of Brown's criterion is that:

> *M* is superior to *m* only if the number of historical episodes for which *T* and *N* coincide is greater than the number of historical episodes for which *t* and *n* coincide.

The second part of Brown's criterion is a principle of coherence: 'that methodology is best . . . which best coheres with other accepted theories'.[75]

Addition of the coherence clause tempers the Historicist emphasis. Reference to historical episodes alone cannot establish

---

[73] G. H. Merrill, 'Moderate Historicism and the Empirical Sense of "Good Science"', *PSA 1980* 1, ed. P. D. Asquith and R. N. Giere (East Lansing: Philosophy of Science Association, 1980), 223.

[74] James R. Brown, 'History and the Norms of Science', *PSA 1980* 1. 241.

[75] Ibid., 241.

the comparative superiority of a methodology. The methodology must also cohere better with 'other accepted theories'. Brown did not specify in detail the types of theories with respect to which coherence is required. However, he did maintain that the coherence requirement would disqualify any methodology whose theoretical and normative interpretations coincide for every historical episode. Such a methodology would triumph over all competitors upon application of the first part of the criterion, but would be disqualified because it contradicts 'prevailing psycho-social theories' according to which the attainment of complete rationality in human affairs is precluded. Brown declared that

a methodology *M* will be obliged to make the likes of the Lysenko affair come out irrational; that is, the theoretical and the normative reconstructions of the Lysenko episode will have to diverge when *M* is applied to this event in the history of science, if the episode is to count in *M*'s favor.[76]

Application of the coherence clause evidently requires reference to 'prevailing psycho-social theories'. Brown left unclear what is comprised under 'other accepted theories'. Is it important that methodologies be consistent as well with prevailing theories of political organization or prevailing theories of religious behaviour? If so, one might appeal to the principle of coherence to support Lysenko–Era genetics. The methodology, application of whose standards justified the Lysenko research programme, conformed to a theory of political organization accepted in the Soviet Union at the time. The methodology, application of whose standards justified the Mendelian research programme in the Soviet Union, did not. If the two methodologies received the same score upon application of the first part of Brown's criterion (admittedly unlikely), then that methodology consistent with Soviet political theory would qualify as the superior methodology. Of course, the phrase 'prevailing theory of political organization' may be taken to refer, as well, to positions held outside the Soviet Union. Application of the coherence clause requires both judgements about the kinds of 'other accepted theories' to be taken into account, and decisions about the extent of the context within which theories 'prevail'. The coherence clause is too vague to be of much use as a criterion of comparative acceptability of methodologies. There are difficulties as well with the first part of the criterion.

[76] Ibid., 242.

M may gain advantage over *m*, not as a result of intrinsic merit, but because *M* subdivides historical developments into more episodes than does *m*. Brown was correct to insist that *HS* cannot be formulated without application of the conceptual apparatus of some *PS*. But he failed to note that the individuation of historical episodes itself reflects judgements of importance rendered from the standpoint of some methodological orientation. For example, the Logical Reconstructionist who holds that scientific theories are linguistic entities may produce a 'theoretical reconstruction' in which there is a 'Bohr-Hydrogen-Atom episode' and a 'Bohr-Sommerfeld-Hydrogen-Atom episode'. A Lakatosian, by contrast, is more likely to compress these developments into a single episode that reflects the rise and triumph of the 'Bohr Research Programme'.

Brown's approach appears to avoid the extreme Historicist and Logicist positions, but both parts of his composite criterion are unsatisfactory. Moreover, it is unclear how the Moderate Historicist is to adjudicate disputes that arise when the result of applying part one conflicts with the result of applying part two.

This is a problem for any Moderate Historicist position. Historical considerations are important to the evaluation of competing methodologies, but so also are logical (or philosophical) considerations. Which considerations take precedence?

Merrill has sought to undermine the position of Moderate Historicism by showing that it reduces either to Unqualified Historicism or Unqualified Logicism. This is a serious charge, since Unqualified Historicism and Unqualified Logicism are both unacceptable.

Unqualified Historicism allows no distinction between a rational reconstruction of scientific progress and the actual course of development of science. Merrill noted that a presupposition of Unqualified Historicism is that those developments that count as developments within the history of *science* can somehow be identified. He emphasized that the needed demarcation of *HS* from the history of philosophy, the history of religion, and so on, cannot be achieved by an appeal to historical considerations alone. Unqualified Historicism thus founders on its inability to circumscribe its own subject matter.

Unqualified Logicism also is an unsatisfactory position. If no historical evidence can be relevant to the justification of evaluative standards, then Unqualified Logicism too founders on the problem of demarcation. The Logicist is unable to produce evidence on behalf of the claim that his normative judgements make reference to *science*.

Merrill's complaints about Moderate Historicism may be telescoped into the following argument:[77]

$$
\begin{array}{ll}
\text{1.} \quad \sim R \supset I & \text{1}'. \quad \sim I \supset R \\
\text{2.} \quad I \supset \sim O & \text{2}'. \quad O \supset \sim I \\
\text{3.} \quad \sim O \supset L & \text{3}'. \quad \sim L \supset O \\
\hline
\therefore \quad \sim R \supset L & \therefore \quad \sim L \supset R
\end{array}
$$

where

R = 'Moderate Historicism reduces to Unqualified Historicism',

I = 'There exists within Moderate Historicism a history-independent criterion for identifying "good science" within actual science',

O = 'Appeal to *HS* within Moderate Historicism can override conclusions reached upon application of the history-independent criterion of "good science"', and

L = 'Moderate Historicism reduces to Unqualified Logicism'.

Premisses 1 and 3 merely unpack the meaning of 'Moderate Historicism'. The status of premiss 2 is less obvious. It is false if conclusions about 'good science' reached upon application of the history-independent criterion can be overridden by appeal to *HS*. If *I* is true then it is possible at least that some episode recorded in a *HS* qualifies as 'good science'. Suppose episode *E*, recorded in some *HS* qualifies as 'good science' upon application of a history-independent criterion. Does it follow that no appeal to *HS* can override the identification of *E* as an episode of 'good science'?

One might appeal to a second *HS* within which the events of *E* are differently interpreted so as to constitute episode *E**. Suppose *E** does not qualify as 'good science' upon application of the history-independent criterion. Is this a case of historical considerations overriding a conclusion reached upon application of the history-independent criterion? Clearly not. If *E* took place, it would still count as 'good science'. The appeal to a second *HS* has shown only that an episode initially believed to have taken place did not take place.

Perhaps an appeal can be made to *HS* other than to redescribe

[77] Merrill, 'Moderate Historicism and the Empirical Sense of "Good Science"', 224–6.

the episode in question. Suppose that there is no dispute over the description of episode E, but that nevertheless there are historical considerations that exclude E from the domain of 'good science'. I don't know what these considerations might be, but if they exclude E from the domain of 'good science', then the history-independent 'criterion' would seem not to be a criterion at all.

It is the Moderate Historicist position that it is neither history-independent criteria alone nor historical considerations alone that determine correct methodological practice, but rather some balance between them. However, the problems which arose about the status of history-independent criteria arise again about the balance selected. Either the appropriate balance is fixed by some history-independent formula or it is free to develop within HS. If the balance is fixed by a history-independent formula then the earlier question may be reinstated—'does it follow that no appeal to HS can override the identification of E as an episode of "good science" by reference to the balance?' Appeal to HS cannot both override conclusions drawn by application of a history-independent balance and preserve the criterial status of the balance. On the other hand, if the balance is determined by historical considerations then application of history-independent criteria cannot override historical considerations, and Moderate Historicism reduces to Unqualified Historicism.

Thus far, Merrill's argument against Moderate Historicism appears to be effective. However, there are versions of Moderate Historicism for which premiss 2 is false and the reduction to Unqualified Historicism or Unqualified Logicism fails. Lakatos's Methodology of Historiographical Research Programmes is one such version. It is Lakatos's position that an appeal to HS can override conclusions reached by application of the history-independent criterion 'incorporation with corroborated excess content'. At any point in time, a new methodology may be formulated whose rational reconstruction of scientific progress incorporates all those episodes rendered rational by the best prior methodology, and more episodes besides. Of course, the HS involved in the override is an HS informed by the canons of the new methodology. Another version is a 'Linear Progress' view according to which the historical development of science—taken as a whole—is progressive. The history-independent criterion is that

evaluative criterion C_2 is superior to C_1
if, and only if,
C_2 informs episodes in science more recent than those episodes informed by C_1.

Thus the evaluative standards and procedures implicit in today's science are superior to those of yesterday just because they are more recent. It is not that yesterday's evaluative practice led to 'bad science', but just that there is a continual linear improvement in such practice.

On the 'Linear Progress' version of Moderate Historicism, premiss 2 of Merrill's argument against Moderate Historicism is false. There is a history-independent criterion, but applications of this criterion are subject to override on behalf of developments within *HS*. As evaluative practices change so too do judgements about 'proper' evaluative practice.

The Linear Progress view is not without its problems, however. Identification of the evaluative standards, application of which contributed to the content of today's science, is itself an interpretation of the historical record. It is a 'theoretical reconstruction' of recent developments, and, as such, it is informed by the conceptual apparatus of some methodology. The suspicion arises that what the methodologist discovers is influenced strongly by what he antecedently is prepared to discover.

But even if there were available a methodologically neutral *HS*, there would remain severe restrictions on the content of the 'linear development' version of Historicism. Certain types of judgement simply cannot be made. It cannot be claimed, for example, that although the actual course of historical development was from evaluative practice P_1 to P_2 to P_3, a more rational progression would have been from P_1 to P_4 to P_5. No appeal is possible from the actual course of evaluative practice (as recorded in some 'standard *HS*').

But how is the 'standard *HS*' to be selected? Suppose Historicist philosophers of science present conflicting versions of the developments that have led to present-day practice. It is then likely that they will also present conflicting views about the evaluative practice which has contributed to the content of present-day science. For instance, one methodologist may conclude that the shelving of an anomaly has contributed to the present acceptance and application of a particular scientific research programme. A second methodologist may disagree on the grounds that there was nothing in the earlier stages of development of the research programme that counts as an anomaly. A 'linear development' version of Historicism justifies the evaluative practice implicit in current science. *But* which interpretation of recent developments is to be selected as the standard *HS*? A crucial weakness of this version of Moderate Historicism is that it can provide no answer to this question.

III EVALUATION PROCEDURES FOR THE JUSTIFICATION OF EVALUATIVE STANDARDS

One promising response to the inconclusive competition at Level 2 among various versions of Logicism and Historicism is to locate the inviolable principle required by prescriptive *PS*, not at Level 2 of the Justificatory Hierarchy, but at a higher level. Suppose, for instance, that there is an inviolable Level 3 procedure for appraising Level 2 justifications. That there does exist such a procedure is a claim advanced by Larry Laudan in *Progress and its Problems*.[78]

Laudan's evaluation procedure

Laudan's proposal is to single out a set of indisputably progressive developments in science and to judge competing reconstructions of scientific progress on their ability to reconstruct these standard cases. Identification of the standard cases is a descriptive undertaking, presumably accomplished upon appeal to the intuitions of the 'scientific élite'. Laudan emphasized the 'pre-analytic' character of the 'preferred intuitions' that establish the standard cases of scientific rationality.[79] These 'preferred intuitions' are the starting-point of the appraisal process and are not themselves subject to evaluation. Given a set of standard cases, competing theories of scientific progress are graded on their ability to reconstruct them. The victorious rational reconstruction is then adopted for the interpretation of historical episodes other than the standard cases.

Laudan's evaluation procedure is not circular. But it does

[78] Larry Laudan, *Progress and Its Problems* (Berkeley: University of California Press, 1977), 155–70.
[79] Ibid., 160–3.

emphasize the interdependence of *PS* and *HS*. It is *HS* that is the source of our intuitions about scientific progress. *PS* is a second-order commentary which sets forth the rational ideal embodied in these intuitions. Hence *PS* is dependent upon *HS* for its subject matter. But *HS* also is dependent upon *PS*. The history of episodes other than the standard cases is a reconstruction based on the rational ideal set forth in a *PS*.

So far as I know, Laudan has not surveyed the 'preferred intuitions' of a properly selected 'scientific élite'. However, he has suggested that such a panel might select as standard cases that

(1) it was rational to accept Newtonian mechanics and reject Aristotelian mechanics by, say, 1800;

(2) it was rational for physicians to reject homeopathy and to accept the tradition of pharmacological medicine by, say, 1900;

(3) it was rational by 1890 to reject the view that heat was a fluid;

(4) it was irrational after 1920 to believe that the chemical atom had no parts;

(5) it was irrational to believe after 1750 that light moved infinitely fast;

(6) it was rational to accept the general theory of relativity after 1925;

(7) it was irrational after 1830 to accept the biblical chronology as a literal account of earth history.[80]

If Laudan's evaluation procedure is to be effective, then certain conditions must be fulfilled:

1. The scientific élite must be selected without reference to their intuitions about the standard cases of scientific rationality. It would be circular to single out members of an élite on the basis of their 'well-founded' judgements about the history of science, and then to accept their intuitions as 'preferred' just because they are members of this élite.

2. A large majority of the élite must agree about the paradigm cases of scientific rationality. Otherwise the fact that a rational reconstruction fitted these episodes would not count for much.

3. There must be consensus, as well, on the way in which standard-case episodes are to be described. Members of the

---

[80] Ibid., 160.

scientific élite may agree, for instance, that the transition from phlogiston theory to oxygen theory was progressive, but have quite different views about the content of 'phlogiston theory' or 'oxygen theory'. Since there is no antecedently given definitive description of standard-case episodes available, consensus of the élite must include agreement on the appropriate description of these episodes.

4. The standard cases agreed upon must enable a discrimination to be made among competing theories of scientific progress. It is conceivable that members of the élite could agree only on low-level laws that are subject to reconstruction from diverse methodological perspectives. Galileo's law of falling bodies, for example, is subject to reconstruction by Inductivism as an extrapolation from empirical findings, and by Cartesian Rationalism as a deductive unfolding of the concept of uniform acceleration. If all the standard-case episodes were of this type, then Laudan's procedure for the comparative justification of methodologies would be ineffective.

5. Criteria of closeness of fit to the standard cases must be specified and applied consistently to competing theories of scientific growth.

Application of Laudan's Level 3 evaluation procedure may yield different results at different times. But the procedure itself is not subject to change. It is put forward as an inviolable transhistorical procedure for the comparative justification of theories of scientific method.

Laudan's proposal to specify an inviolable transhistorical procedure at Level 3 of the Justificatory Hierarchy has certain advantages. The procedure combines prescriptive and descriptive emphases, thereby avoiding the extremes of historical relativism and a normativism unresponsive to developments in *HS*. Laudan's procedure insures prescriptive status for *PS* in such a way as to allow for the evolution of evaluative standards and theories of scientific progress at lower levels of the Justificatory Hierarchy. In addition, Laudan's evaluation procedure is applicable without restriction to theories of scientific progress. It is not limited, for example, to theories of scientific progress that are related by incorporation.

The above advantages provide no support for Laudan's claim that the principles of his own Problem-Solving Model of scientific progress stipulate the 'general nature of rationality'. According to Laudan, the Problem-Solving methodology

transcends the particularities of the past by insisting that for all times and for all cultures, provided those cultures have a tradition of critical discussion (without which no culture can lay claim to rationality), rationality consists in accepting those research traditions which are most effective problem solvers.[81]

On Laudanian principles, this claim ought be evaluated by reference to the Level 3 evaluation procedure. Which theory of scientific progress is judged best upon application of Laudan's Level 3 evaluation procedure depends, in part, on empirical considerations. These considerations include the determination of the preferred intuitions of members of the scientific élite. Given consensus on standard cases of scientific rationality, one theory of scientific progress may emerge as victor in the evaluation process. But such victory would be provisional. The judgements of members of the scientific élite of year 2000 may be quite different. And if 'preferred intuitions' change, so too does the evaluation of theories of scientific progress.

Other Level 3 evaluation procedures

Laudan's proposal for the evaluation of competing methodologies is not the only option available at Level 3. Another option is to anchor the procedure differently. One might argue that the choice of a group of élite scientists to anchor the procedure is not the best choice. Although one would expect an eminent scientist's judgements about recent developments in his own area of specialization to be well informed, there is less reason to suppose that his judgements about the remote history of his discipline, or the history of other disciplines, are well informed. It might be more appropriate instead to anchor a Level 3 evaluation procedure in the judgements of professional historians of science. On this approach, it is historians of science who are called upon to reach consensus on a set of 'indisputably progressive' episodes. The consensus, if it can be achieved, reflects agreement on what took place, and what was significant about, important scientific episodes. Competing methodologies can then be gauged on the basis of the congruence of their rational reconstructions and the designated set

---

[81] Ibid., 130.

of 'progressive episodes'. The victorious *PS* becomes the source of the basic methodological judgements used in the historical reconstruction of episodes other than the 'indisputably progressive' set.

This choice of an anchor may also be challenged. For instance, it may be objected that what is lacking in an approach anchored in a set of standard cases selected by historians of science is a principle that weights preferentially the most recent developments in science. Does it even matter whether a contemporary methodology can show that selected sixteenth-century developments were rational? One may argue that the best procedure is the procedure which selects that *PS* whose principles conform most closely to the methodological implications of *recent* developments in science.

A Level 3 procedure consistent with this emphasis on recent science might begin with the selection of an élite group of scientists charged with the task of singling out, say, the twenty most important scientific developments of the past twenty years.[82] It is likely that appointment to membership of the group of élite scientists itself would be controversial. Someone would have to do the selecting, and the selector or selectors would have to decide both appropriate group quotas for theoretical physicists, botanists, anthropologists, and so on, and which specific individuals are best qualified.

Suppose widespread agreement were achieved on appointments to the group of élite scientists. For the subsequent evaluation procedure to be effective, there would have to be agreement on the part of a substantial majority of the scientists on both the appropriate description of progressive episodes and the significance of the episodes thus described. Given agreement on a set of 'standard-case' episodes from recent science, competing methodologies could then be ranked on their ability to reconstruct these episodes.

Each Level 3 procedure outlined thus far takes as starting-point the judgements of some élite group. The preferred intuitions of the élite are not subject to criticism. Of course, one may set aside the requirements of a particular procedure in order to examine the appropriateness of its starting-point. But to practice a Laudanian-type procedure at Level 3 is to accept as the first stage of that

---

[82] The number 'twenty' is not privileged. Other numbers may be designated, provided that the episodes that qualify can be defended as 'recent'.

procedure whatever judgements are rendered by the members of the élite group.

Perhaps a Level 3 procedure can be formulated within which *all* stages are open to criticism. Within such a procedure both initial judgements and the reconstructions of these judgements on behalf of competing methodologies would be subject to modification in response to criticism. Paul Thagard has sketched a procedure of this type.[83] He did not recommend its adoption. Nevertheless the procedure is an interesting alternative to procedures whose starting-point is the judgements of the members of an élite group. It is based on John Rawls's concept of 'reflective equilibrium'.[84]

Rawls had suggested that an equilibrium be achieved between our intuitions about justice and the principles of justice that would presumably be selected by rational individuals who pursue self-interest under a 'veil of ignorance' about themselves[85] and their place in society.[86] In Thagard's adaptation of this concept, the equilibrium in question is between methodological principles explicitly or implicitly presupposed in the historical unfolding of a set of standard-case episodes and methodological principles which inform the selection and reconstruction of these episodes. The modified principles that emerge upon the achievement of reflective equilibrium then are taken to be the basic principles of a 'normative model' that provides the appropriate methodological perspective for the reconstruction of historical episodes other than the standard cases.

The starting-point of this procedure is the selection, under some description, of a set of standard-case episodes. A presupposition of the procedure is the existence of a *HS* in which scientific developments are recorded. Principles of selection are then applied to the historical record to single out episodes that exemplify scientific progress. The principles applied are the principles that inform our initial judgements about scientific progress. Thagard recommended the general principle that standard-case episodes be selected 'on the basis of subsequent events in the history of science which have marked the cases as

---

[83] Paul Thagard, 'From the Descriptive to the Normative in Psychology and Logic', *Phil. Sci.* 49 (1982), 24–42.

[84] Ibid., 39.

[85] The veil of ignorance covers the aptitudes, resources, preferences, and concepts of the good of each individual.

[86] John Rawls, *A Theory of Justice* (Oxford: Clarendon Press, 1972), 136–50.

significant contributions to the growth of scientific knowledge'.[87]

To implement this general principle it is necessary to apply some specific measure of historical influence. One such measure is the extent to which a development is cited in the relevant literature. Armed with this indicator one might single out those developments that are acknowledged to have been important to the formulation of present-day theories. One then might select these episodes to be the paradigmatic cases of scientific progress.

Given a set of standard-case episodes, the next step in the procedure is to uncover the evaluative standards explicitly or implicitly presupposed in the historical unfolding of these episodes. The methodological principles presupposed in the unfolding of the standard-case episodes are then compared with the methodological principles assumed initially in the selection of these episodes. Modifications and adjustments are made until a 'reflective equilibrium' is reached. The principles reached by reflective equilibrium then provide the appropriate methodological perspective for the interpretation of historical episodes other than the standard-case episodes.

The procedure thus outlined is a plausible alternative to those procedures which take as starting-point the judgements of an élite group of scientists, historians, or philosophers. An important feature of the procedure under consideration is a process of accommodation that produces a reflective equilibrium.

'Reflective equilibrium' is a balance in which antithetical, or at least divergent, commitments are accommodated.[88] The accommodation is marked by a reduction in the tension initially produced by the discordant emphases. An important question is whether a diminution of tension is in itself a sufficient condition of the achievement of reflective quilibrium. If it is, then reflective equilibrium may be achieved simply by learning to live with the discord. Unless an objective measure of reflective equilibrium can be formulated, this Level 3 procedure rests ultimately on the subjective appraisals of some group of practitioners. The search for reflective equilibrium then would differ from the Laudanian position and its variants only by deferring the appeal to subjective responses to a later stage of the procedure.

---

[87] Thagard, 'From the Descriptive to the Normative', 27.
[88] Rawls, *A Theory of Justice*, 20.

IV JUSTIFICATION OF EVALUATION PROCEDURES FOR
THE SELECTION OF A PHILOSOPHY OF SCIENCE

Given a competition among proposed Level 3 evaluation procedures, there are a number of approaches available to identify the 'best procedure'. To debate the merits of Level 3 procedures is to conduct an evaluation at a still higher level of justification. And one may elect to locate the inviolable principle required for prescriptive status, not at Level 3, but at Level 4 of the Justificatory Hierarchy.

A straightforward approach to the selection of an evaluation procedure would be to conduct a vote of members of the Philosophy of Science Association. The evaluation procedure that received the greatest number of votes would be declared to be 'the best' procedure. And the principle that a vote of Philosophy of Science Association members determines the appropriate Level 3 evaluation procedure may be taken to be the inviolable principle required for prescriptive status.

A second approach to the selection of an evaluation procedure might be to apply at Level 4 the Lakatosian formal criterion 'incorporation with corroborated excess content'. At Level 4, the Lakatosian criterion would be applied to the rational reconstructions of scientific progress implied by those methodologies that are selected by competing Level 3 procedures. Given a set of rational reconstructions, the members of which conform to the relation of incorporation with corroborated excess content, the victorious evaluation procedure is that procedure which selects the methodology which implies the most inclusive and best corroborated rational reconstruction. Of course the Lakatosian criterion is inapplicable to cases of partial overlap, regardless of whether it is applied at Level 2 or Level 4 of the Justificatory Hierarchy.

Other approaches may be devised which are not limited in this way. For instance, the following criterion of comparative justification might be applied at Level 4:

given PS_1 and PS_2 selected by evaluation procedures E_1 and E_2 respectively, E_1 is superior to E_2 provided that the number of standard-case episodes selected as specified by E_1 and reconstructed by PS_1 is greater than the percentage of standard-case episodes selected as specified by E_2 and reconstructed by PS_2.

This criterion may be taken to be an inviolable principle. It is not without drawbacks, however. The number of standard-case episodes delineated by a methodology may reflect nothing more than an inessential preference for 'lumping' or 'splitting'. It would be inappropriate to reward one methodology solely because it subdivides into three historical episodes what a second methodology has reconstructed as a single three-stage episode.

Additional criteria of Level 3 principles may be formulated. The above survey is certainly not exhaustive. But since each principle put forward thus far has proved open to objections, it is appropriate at this point to give serious consideration to an alternative view of *PS*.

7
Philosophy of Science Without Prescriptive Intent

THE foregoing examination of candidates for inviolable status at the various levels of the Justificatory Hierarchy is based on the assumption that *PS* is a prescriptive discipline. If the examination has been conducted fairly, there are grounds for pessimism about the availability of principles worthy of inviolable status.

Perhaps it is a mistake to insist upon prescriptive status for *PS*. Perhaps the work of the philosopher of science should be restricted to an uncovering of the methodological standards and procedures that have informed scientific practice. On this understanding of his role, the philosopher of science refrains from issuing recommendations about 'proper' evaluative practice. Instead he displays the methodological standards and procedures that have been utilized during various episodes from *HS* (including recent science). Needless to say, no evaluative principle is held to be inviolable within such a *PS*. What can be said in favour of this alternative?

Certainly the descriptive alternative has the virtue of modesty. There is something pretentious about a *PS* which purports to educate scientists about proper evaluative practice. No such prescription is countenanced on the descriptive alternative. Instead, the philosopher of science remains neutral. He is an exhibitor and not an advocate. Contemporary scientists are left free to apply, modify, or ignore the standards uncovered by philosophers of science.

Then too the descriptive alternative may derive support from a failure of prescriptive *PS* to engage the attention of scientists. Gerald Holton has argued that the epistemological debates important to the development of early twentieth-century science no longer fuel scientific progress.[1] Scientists such as Bohr,

---

[1] Gerald Holton, 'Do Scientists Need a Philosophy?' *Times Literary Supplement*, 2 Nov. 1984, 1231–4.

Einstein, Bridgman, Heisenberg, de Broglie, and Poincaré had been greatly concerned about the epistemological implications of simultaneity, indeterminacy, discontinuity, and wave–particle dualism. Moreover, *PS*, as expressed by Hume, Mach, Hertz, Pearson, and so on, was an important part of the educational background of these men. According to Holton, the situation has changed (at least among theoretical physicists). He declared that it is the 'perception by the large majority of scientists, right or wrong, that the messages of more recent philosophers, who themselves were not active scientists, are essentially impotent in use, and therefore may be safely neglected'.[2] If *PS* is to be judged on the basis of its impact upon scientific practice, then Holton's conclusion, if correct, provides indirect support for the descriptive alternative. It does so because the conclusion is a negative judgement on the value of prescriptive *PS*.

The descriptive alternative does not reduce the philosopher of science to a mere chronicler of professed methodological commitments. He is charged to uncover the methodological principles actually effective within scientific practice. These principles may or may not be affirmed explicitly by the scientists in question. In some cases the methodological principles to which scientists pay homage are not the principles which inform their work. For example, when Darwin declared that 'I worked on true Baconian principles, and without any theory collected facts on a wholesale scale',[3] it is a task for the philosopher of science to judge whether the available records—papers, letters, journal entries, and so on—support this claim. And when Newton claimed that the laws of motion and gravitation were discovered by 'inferring particular propositions from the phenomena' and 'rendering general these propositions by induction',[4] it again is a task for the philosopher of science to judge whether these claims made for 'inductive inference' are supported by the available historical evidence.[5]

[2] Ibid., 1232.

[3] Charles Darwin, *Autobiography*, in *The Life and Letters of Charles Darwin* 1, ed. Francis Darwin (New York: Basic Books, 1959), 68.

[4] Isaac Newton, *Mathematical Principles of Natural Philosophy*, trans. A. Motte, rev. F. Cajori (Berkeley: University of California Press, 1962), 2. 547.

[5] See John Losee, *A Historical Introduction to the Philosophy of Science* (Oxford: Oxford University Press, 1980), 81–90.

II ANTICIPATIONS OF DESCRIPTIVE PHILOSOPHY OF SCIENCE

Anticipations of a descriptive approach to *PS* may be found in the writings of N. R. Hanson, Stephen Toulmin, Paul Feyerabend, Dudley Shapere, and Larry Laudan.

Hanson on the uses of scientific laws

N. R. Hanson outlined a programme for a philosophy of physics in which the philosopher exhibits the various uses to which scientific laws are put. As a contribution to this programme he set forth the various uses of Newton's Second Law within mechanics.[6]

One use is to take the Second Law as providing an explicit definition of 'force', such that 'force' may be replaced in each of its occurrences by the phrase 'mass times acceleration'. If this replacement is carried out systematically, the term 'force' is eliminated from dynamics. This is a position that was advocated by Boltzmann, Kirchhoff, and Mach. They believed that the concept of force should be eliminated from physics because of its anthropomorphic overtones, which result from association of the concept with experiences of muscular exertion.

At the other extreme, Newton's Second Law has been tested as if it were an empirically significant generalization. George Atwood designed a machine to confirm the law. He attached unequal weights to the ends of a silk thread, passed the thread over a pulley, and measured the resulting acceleration of the weights. He compared observed accelerations with accelerations calculated from the Second Law and found good agreement. Atwood certainly believed that he had confirmed an empirically significant generalization, despite the fact that if the observed accelerations had been systematically higher or lower than the calculated values, agreement could have been restored by changing the value of 'g', the earth's gravitational constant.

There are other ways to treat Newton's Second Law as an empirically significant generalization. One such approach is to measure accelerations produced by Hooke's Law forces. Since Newton's Law implies a direct proportionality of force and

---

[6] N. R. Hanson, *Patterns of Discovery* (Cambridge: Cambridge University Press, 1958), 99–118.

acceleration (provided that the mass is constant), the independent determination of forces and corresponding accelerations may be cited as confirming evidence for the Law. It is clear that Newton's Second Law is used by physicists as a generalization subject to confirmation by experimental evidence.

Newton's Second Law is also used as a methodological rule to guide research in dynamics. The Law stipulates that an observed acceleration is to be correlated with some force. It is the physicist's task to specify the nature of this force. Typically, the physicist specifies the nature of a force by stating the way in which the force depends on such factors as distance, velocity, and physical constants.

A failure to correlate successfully a particular force-function with an observed acceleration does not falsify Newton's Second Law. In its use as a guide to research, the Law would be abandoned only if it could be shown that no possible force-function, however complex, has the required Second Law relation to the observed acceleration. Such a demonstration is impossible in principle.

Still another use of Newton's Second Law is as a rule of inference. Given information about the state of a dynamical system in one region of space and time, the Law may be used to make predictions (or retrodictions) about the system in some other region of space and/or time. Just as *modus ponens* and *modus tollens* specify ways in which conclusions may be derived from premises, so also the Second Law specifies how successive states of a system may be derived from information about its initial state. As a rule of inference, the Second Law is subject to neither confirmation nor falsification. Any discrepancies between observed motions and inferences drawn in accordance with the Law are attributed to incorrect specification of one or more of the states of the system.

Hanson emphasized that 'what physicists call "the second law" really consists in everything that can be expressed by way of different uses of this formula'.[7] He pointed out that the uses sketched above, and others besides, have contemporary applications. It is not the case that the law was once considered to be an empirical generalization but that it is now used exclusively as an a priori principle.

[7] Ibid., 105.

Hanson sought to restrict the role of the philosopher of physics to an uncovering of the diverse uses of scientific laws in the solving of physical problems. He declared that

to go further—to make philosophical recommendations about the *real* nature of the laws of dynamics or about how law formulae *ought* to be used—is exactly what the theoretical physicist is trained for. The philosopher of physics who ventures into this territory must expect to be judged by standards unknown in the British Academy.[8]

Hanson conceded that prescriptive judgements are important within physics. However, he noted that it is scientists such as Kepler, Galileo, Newton, Maxwell, Einstein, Bohr, Schrödinger, and Heisenberg who have established the permitted functions of the laws of physics.[9] The philosopher is out of his depth when he seeks to legislate proper evaluative procedures for physics.

Toulmin on Ideals of Natural Order

Stephen Toulmin's discussion of 'Ideals of Natural Order' (1961)[10] emphasized the role of basic assumptions in the development of science. According to Toulmin, Ideals of Natural Order are standards of regularity that 'mark off for us those happenings in the world around us which do require explanation by contrasting them with "the natural course of events"—i.e., those events which do not'.[11]

Newton's first law is such an ideal. It specifies that uniform rectilinear motion is inertial motion, and that it is only changes in such motion that need to be explained. Newton's ideal of natural order displaced a corresponding Aristotelian ideal. Aristotle had taken as the paradigm case of local motion the dragging of a body over a resisting surface. The speed reached by such a body depends on the ratio of the effort exerted to the resistance offered. The very presence of motion indicates that an effort is being applied. On the Aristotelian ideal of natural order, it is motion itself that needs to be explained and not just changes of motion.

Commitment to Ideals of Natural Order affect scientists' expectations. And it is because of these expectations that scientists

---

[8] Ibid., 113. [9] Ibid., 113.

[10] Stephen Toulmin, *Foresight and Understanding* (New York: Harper Torchbooks, 1961), 44–82.

[11] Ibid., 79.

come to regard certain phenomena as anomalies in need of explanation. In the case of the Aristotelian ideal, the continuing motion of a javelin after the hurler has released it demands an explanation. But the airborne javelin appears to be subject to no effort. Aristotle suggested, with some hesitation, that the successively adjacent air transmits to the projectile a propensity to continue in motion.[12] Needless to say Aristotelian natural philosophers were uneasy about 'explanations' of this type. Toulmin suggested that the recognition of anomalies has provided a stimulus for the creation of new Ideals of Natural Order.

Toulmin did not issue prescriptive recommendations on behalf of specific Ideals of Natural Order. Instead he sought to develop an explanatory model to account for the historical development of these ideals. The explanatory model invokes an analogy to the Theory of Organic Evolution. Toulmin maintained that the identity-through-change of a scientific discipline is analogous to the identity-through-change of a biological species (see Table 7).[13]

Table 7. Toulmin's Evolutionary Analogy

	Organic evolution	Conceptual change
Unit of study:	Species	Scientific discipline
Comprised of:	Individual organisms	Concepts, methods, and aims
Units of variation:	Mutant forms within the population at t_1	Conceptual variants within the discipline at t_1
Units of effective modification:	Those t_1 variants dominant within the population at t_2	Those t_1 variants dominant within the discipline at t_2
Mechanism of selection:	Differential reproductive pressure	Need for deeper understanding

Toulmin held that conceptual development within a scientific discipline is a 'natural selection' which operates on a set of 'conceptual variants'. The rational reconstruction of scientific growth is thus an analysis of 'intellectual ecology over time'. As such, the philosophy of science is related to the history of science

---

[12] Aristotle, *Physics*, Book VII, 267a.
[13] Toulmin, *Human Understanding* (Oxford: Clarendon Press, 1972), I. 121–3; 135–44.

Table 8. Evolution of Species and Evolution of Concepts

	Evolution of species		Conceptual change	
	Phylogeny	Ecology	History of science	Philosophy of science
Question	From what succession of precursors has this species descended?	By what sequence of responses to environmental pressures did the species acquire its present form?	From what succession of precursor concepts has this set of concepts descended?	By what sequence of responses to disciplinary pressures did this set of concepts arise?
Answer	A tree of descent	Application of Theory of Natural Selection	A history of a scientific discipline	A rational reconstruction of scientific growth

much as ecology is related to phylogeny in evolutionary biology (see Table 8).[14]

On this understanding of *PS*, its basic questions are historical questions.

Toulmin sensed that practitioners of prescriptive *PS* would be dissatisfied with his version of *PS*:

> But (some will want to ask) can we not at least abstract certain general and universal criteria of scientific merit from the diversity of historical contexts; and so create a secure base, or forum, from which we can pass philosophical judgment on scientific changes 'from outside' the chances of the historical process?[15]

Toulmin maintained that the attempt to formulate inviolable evaluative standards for science is misguided. It is a matter of historical record that evaluative procedures evolve. Standards held to be inviolable at one point in the development of science are qualified or abandoned subsequently. Given the history of evaluative practice it is both futile and pretentious to prescribe context-independent criteria of 'fitness'.

Whewell and Toulmin each proposed a 'theoretical reconstruction' of *HS* on behalf of a conceptual model. Whewell's interpretive

---

[14] Toulmin, 'Rationality and Scientific Discovery' in *Boston Studies in the Philosophy of Science* 20, ed. K. Schaffner and R. S. Cohen (Dordrecht: D. Reidel, 1974), 402–3.
[15] Ibid., 403.

model features the Aristotelian concept of a set of distinct sciences, each with appropriate basic predicates and first principles, and a 'Kantian' polarity of fact and idea. Toulmin's interpretive model features the conceptual structure of the Theory of Organic Evolution. But whereas Whewell sought to read from his reconstruction inviolable evaluative principles, Toulmin did not.[16] I suspect that this is because there is an important difference in the nature of the models they apply.

A presupposition of Whewell's model is that there exist immutable relations among the fundamental ideas of a science and that knowledge of these relations may be achieved within *HS*. It is consistent with these assumptions that there exist inviolable standards such as consilience.

Toulmin presupposes only that an analogue of natural selection operates on conceptual variants within *HS*. On Toulmin's model, it is the 'fit' theories that survive, but 'fitness' is always context-relative. Moreover, a present adaptation to 'ecological' pressures is successful only if a balance is struck between adaptation to present conditions and retention of capacity to respond creatively to future changes of these conditions. To take seriously the evolutionary analogue is to emphasize the importance of 'ecological considerations' in conceptual development. A conceptual system may continue to be successful because of its flexibility in the face of changing intellectual conditions. But success may also be achieved by a non-flexible conceptual system just in case the 'ecological environment' remains the same. Since 'success' is a matter of future adaptability as well as present adaptation, the only context-independent evaluative standard available at Level 1 of the Justificatory Hierarchy is the requirement that a balance be achieved between present adaptation and future adaptability. The evolutionary model and descriptive *PS* are well matched.

---

[16] Toulmin did claim to be recommending an 'impartial, objective' standpoint. He declared that 'there is one basis, and one alone, on which our judgments of "rationality" and conceptual "merit" can truly be impartial. This is one that takes into account the experience which men have accumulated when dealing with the relevant aspects of human life—explanatory or judicial, medical or technological—in *all* cultures and historical periods' (*Human Understanding*, 1. 500). Such a pronouncement is too vague to be of use as an evaluative principle. Moreover, if it is taken to be a straightforward appeal to inclusiveness, then it is inadequate. Synoptic vision is not achieved simply by collecting various particular perspectives. Judgements of importance are required. What is needed is some way to distinguish an adequate from an inadequate synthesis of diverse perspectives.

Feyerabend on methodological anarchism

Paul Feyerabend has issued recommendations which at first glance appear to support the programme of descriptive *PS*. What philosophers of science ought do, he suggested in a 1970 essay, is to 'return to the sources'. Philosophers have constructed 'beautiful but useless formal castles in the air'. These castles ought be replaced by a 'detailed study of primary sources in the history of science'.[17]

Subsequently, Feyerabend has promoted a 'methodological anarchism'. In *Against Method* (1975), he maintained that 'the idea that science can, and should, be run according to fixed and universal rules, is both unrealistic and pernicious'.[18] Feyerabend sought to demonstrate that certain generally acknowledged cases of scientific progress are inconsistent with both the 'best' methodology of that time and presently accepted methodologies. Of course, this does *not* prove that science *cannot* 'be run according to fixed and universal rules'. Ultimately, Feyerabend rests his case on the unproved assumption that commitment to inflexible standards makes science 'less adaptable and more dogmatic'.[19] Or rather, such commitment to inflexible standards *would* thwart progress in science *if* scientists paid any attention to advice from philosophers of science.

It has been argued above (in Chapter 5) that every prescriptive *PS* contains at least one evaluative principle that is held to be inviolable. Feyerabend has maintained that commitment to inviolable standards is inimical to scientific progress. A repudiation of the programme of prescriptive *PS* would be consistent with 'methodological anarchism'.

However, Feyerabend has instead superimposed some prescriptive recommendations upon his anarchistic manifesto. He noted that the converse side of the denial that there should exist transhistorical evaluative standards is the methodological directive 'anything goes'.[20] In itself, 'anything goes' is an innocuous methodological principle. But upon occasion Feyerabend has put forward this

---

[17] Paul Feyerabend, 'Philosophy of Science: A Subject with a Great Past' in *Historical and Philosophical Perspectives of Science*, ed. R. Stuewer (Minneapolis: University of Minnesota Press, 1970), 183.

[18] Feyerabend, *Against Method* (London: NLB, 1975), 295.

[19] Ibid., 295.

[20] Ibid., 23.

directive principle in a more specific form as a principle of theory proliferation.

According to Feyerabend, the proper approach to a domain in which one theory reigns supreme is to invent alternative theories. He noted that, even if an alternative theory fails to replace the established theory, it may contribute to progress by suggesting additional tests of the entrenched theory. Often a really pertinent test of a theory becomes available only after a competing theory has been formulated.[21] Feyerabend also recommended the creation of theories that contradict 'well-established facts'. Facts often gain their status on the basis of concealed assumptions, and the invention of rival theories may bring into question these assumptions.[22]

That Feyerabend tempers his methodological anarchism with specific prescriptive recommendations is not a sign of wavering intention, given his stated objective of conforming to the precepts of Dadaism.[23] One achieves a Dadaist stance only upon recognition that one must also assume an anti-Dadaist stance.[24] If Feyerabend succeeds in his apprenticeship to Dadaism, efforts to classify his work are bound to fail.[25]

The interpreter of Feyerabend's work faces the further problem that he is in favour of obliterating disciplinary boundary lines. In particular he defends a position which, from one perspective, recommends expansion of *PS* toward the broader concerns of cultural development, and, from an equally valid perspective, recommends the disappearance of *PS* as a distinct discipline.

Shapere on the development of scientific domains

Shapere's programme for a non-presuppositionist, but prescriptive, *PS* was criticized in Chapter 4. Despite the fact that Shapere sought to formulate a prescriptive *PS*, his analysis of the development of scientific domains anticipates the descriptive approach to *PS*. The concept of a 'domain' is a category for the

[21] Feyerabend, 'Against Method: Outline of an Anarchistic theory of Knowledge' in *Minnesota Studies in the Philosophy of Science* 4, ed. M. Radner and S. Winokur (Minneapolis: University of Minnesota Press, 1970), 26.

[22] Feyerabend, 'Problems of Empiricism, Part II' in *The Nature and Function of Scientific Theories*, ed. R. Colodny (Pittsburgh): University of Pittsburgh Press, 1970), 307–23.

[23] Feyerabend, *Against Method*, 21 n., 33 n., 189, 191.

[24] Ibid., 189.

[25] In *Science in a Free Society* ((London: NLB, 1978), 125–217), Feyerabend has complained extensively about misrepresentations of his views.

interpretation of the history of science. A domain is a body of information which comprises a unified subject matter. In a domain items of information are associated such that:

(1) The association is based on some relationship between the items;
(2) There is something problematic about the body so related;
(3) That problem is an important one; [and]
(4) Science is 'ready' to deal with one problem.[26]

Shapere noted that two sorts of problems may arise with respect to a domain. There may be a problem of classification, requiring decisions to be made about borderline cases. Such a problem is a 'domain problem' and its solution results in a clarification of the domain itself. But there may also be a 'theoretical problem', the solution of which would be a theory about why the items of information do constitute a domain. For instance, a theoretical problem might be solved by specifying a microtheory about the structure of the domain-objects.[27]

Shapere has contributed a number of case-studies of domain development. One such domain is the periodically-ordered chemical elements. According to Shapere, this domain exhibits a 'Principle of Discrete Compositional Reasoning', namely 'to the extent that a domain D satisfies the following conditions or some subset thereof, it is reasonable to expect (or demand) that a discrete compositional theory be sought for D'.[28] Among the conditions specified are that D is ordered, that the order is periodic, and that domain-items have values that are integral multiples of a fundamental value.

A second domain studied by Shapere is the spectral classification of stars. Astronomers working at the close of the nineteenth century increasingly came to interpret the spectral pattern of a star as an indicator of its age (blue-white in youth, red in old age). This domain exhibits a 'Principle of Evolutionary Reasoning', in which a discrete ordering of the properties of domain-objects is correlated with a sequential temporal ordering.[29]

[26] Dudley Shapere, 'Scientific Theories and Their Domains' in *The Structure of Scientific Theories*, ed. F. Suppe (Urbana: University of Illinois Press, 1974), 525.
[27] Ibid., 533–4.
[28] Shapere, 'Discovery, Rationality, and Progress in Science: A Perspective in the Philosophy of Science' in *Boston Studies in the Philosophy of Science* 20, ed. K. Schaffner and R. S. Cohen (Dordrecht: D. Reidel, 1974), 409.
[29] Shapere, 'Scientific Theories and Their Domains', 549–55.

Thus far Shapere's enquiry has been restricted to an uncovering of methodological principles that have informed episodes within *HS*. However, after having exhibited these methodological principles, he proceeded to recommend them as principles directive of further research. If a domain satisfies specified conditions, then scientists *ought* to formulate theories which conform to the appropriate Principle of Reasoning. This, of course, is to move from a generalization about scientific practice to a prescription for achieving 'good science'. Although Shapere's investigation of the development of domains anticipates a descriptive approach to *PS*, the *PS* which he eventually recommended does contain explicit prescriptive recommendations.[30]

Laudan's Reticulational Model of Justification

In *Progress and Its Problems* (1977), Laudan had developed a Level 3 Procedure for the evaluation of alternative philosophies of science. The procedure is anchored by the 'preferred intuitions' of a scientific élite. It is applicable to rational reconstructions of scientific progress at Level 2 of the Justificatory Hierarchy. Included among these Level 2 reconstructions is the reconstruction associated with Laudan's own Problem-Solving Model of scientific progress.

Subsequently in *Science and Values* (1984),[31] Laudan criticized the hierarchical model of justification which he had applied in *Progress and Its Problems*. The 'hierarchical justificatory ladder' has three rungs: (1) laws and theories, (2) methodological rules, and (3) claims about the 'basic cognitive aims' of science (the axiological level). He noted that the hierarchical justificatory ladder is often affirmed in conjunction with a commitment to a 'Leibnizian Ideal' that disputes at one level are resolved by application of principles at the next higher level.[32] Disputes about laws and theories are resolved upon appeal to the methodological level and disputes about methodological principles are resolved upon appeal to the axiological level.

Laudan maintained that these assumptions are at variance with actual evaluative practice within *HS*. In the first place, historical

[30] See Chapter 5, pp. 65–71.
[31] Larry Laudan, *Science and Values* (Berkeley: University of California Press, 1984).
[32] Ibid., 5–8; 23–6.

enquiry reveals a measure of underdetermination within the hierarchy. Some, but by no means all, disputes at the theoretical level are resolved by application of methodological rules, and some, but by no means all, disputes about methodological rules are resolved by reference to shared aims. Moreover, consensus about methodological rules is neither sufficient nor necessary for consensus about laws and theories, and axiological consensus is neither sufficient nor necessary for consensus about methodological rules. In addition, axiological disputes themselves are not subject to resolution by appeal to a higher level of analysis. Since there is no further rung to the ladder, no procedure is available for the adjudication of disputes about the basic aims and goals of science. Laudan noted that within Popper's version of the hierarchical model, for example, 'there was no definitive or rational way to choose between [internally consistent versions of] realism and instrumentalism'.[33]

Laudan emphasized that, on the hierarchical model, there is no provision for modification of axiological principles by appeal to considerations at the lower levels. In this respect the hierarchical model fails to fit important episodes from *HS*. Laudan called attention to the tension within late eighteenth-century science between the acknowledged goals of Newtonian 'Experimental Philosophy' and a variety of theories which postulated the existence of unobservable entities. Newton had proposed, and many eighteenth-century scientists had affirmed, that 'Experimental Philosophy' be restricted to statements about 'manifest qualities', 'theories' induced from these statements, and queries directive of further enquiry. Among the theories which postulated hypothetical entities were Boscovich's matter theory about dimensionless point-masses subject to a complex force–distance law that oscillates between attraction and repulsion, the phlogiston theory which interpreted combustion to be a process in which a substance is exuded from the burning material, Hartley's neurological theory about the action of ethereal fluids, Franklin's one-fluid theory of electricity, and Lesage's theory of gravitational corpuscles. Laudan pointed out that 'what confronted all these scientists was a manifest conflict between the "official" aims and goals of science and the types of theories they were constructing'.[34]

---

[33] Ibid., 48.   [34] Ibid., 57.

The tension between aims and theories was resolved in the nineteenth century by a revision of the axiological level to legitimate the postulation of hypothetical entities. Since the hierarchical model does not accommodate those interactions in which the axiological level is modified from below, Laudan concluded that 'the pecking order implicit in the hierarchical approach must give way to a kind of leveling principle that emphasizes the patterns of mutual dependence between these various levels'.[35]

The hierarchical model has been superimposed upon evaluative practice within science and found not to fit. Laudan suggested that the hierarchical model be replaced by a Reticulational Model that does justice to the reciprocal relations between theories and methodological rules, methodological rules and cognitive aims, and cognitive aims and theories (see Figure 6).[36]

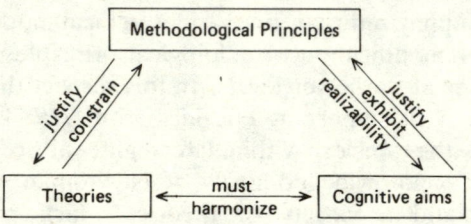

Fig. 6. Laudan's Reticulational Model of Justification

Laudan conceded that the Reticulational Model provides only weak constraints on the selection of cognitive aims for science. Diverse aims may coexist and participate in the appropriate relations to theories and methodological principles. Laudan emphasized that no single aim, methodological rule, or theory has the status of an inviolable principle within the model. He echoed Shapere's claim that all evaluative principles are subject to change within science.

The triadic network is not itself an evaluative principle. One cannot apply the Reticulational Model to justify particular methodological rules or cognitive aims. However, one can apply the model to criticize interpretations of scientific evaluative

<hr />

[35] Ibid., 63.   [36] Ibid., 63.

practice that fail to emphasize the dynamic equilibria singled out by the model.

Laudan assigned normative status to the Reticulational Model in both descriptive and prescriptive senses. In its descriptive usage, the model sets forth 'the way in which debate about cognitive values in science is actually conducted'.[37] In its prescriptive usage, the model 'purports to give a normatively viable characterization of how discussions about the nature of cognitive values should be conducted'.[38] The PS of *Science and Values* thus retains prescriptive intent. However, its normative–prescriptive analysis is directed, not at rational reconstructions of scientific progress, but at philosophical theories about the cognitive aims of science.

Laudan's principle prescriptive application of the Reticulational Model is a criticism of certain doctrines associated with the position of epistemic realism.[39] Among these doctrines are the methodological principle of 'inference to the best explanation', and the limiting-case view of theory replacement according to which acceptable new theories ought to retain 'significant portions of the theoretical content (or extension) of their successful predecessors'.[40]

Laudan maintained that these methodological principles do not stand in the appropriate relationship to the professed cognitive aims of realism. The Reticulational Model requires that reflective equilibrium be achieved between methodological rules and cognitive aims, and, according to Laudan, there is disharmony rather than equilibrium within the realist interpretation of science. In addition, Laudan pointed to an incongruity between the cognitive aims of realism and certain successful and important scientific theories. The theories in question invoke concepts that lack reference and thus fail to satisfy the realist ideal.[41].

It is beyond the scope of this study to assess the adequacy of Laudan's criticism of epistemic realism. What is important in the

[37] Laudan, *Science and Values*, 104.
[38] Ibid., 104.
[39] Laudan characterized 'epistemic realism' as 'the claim that certain forms of evidence or empirical support are so epistemically probative that any theory that exhibits them can legitimately be presumed to be true, or nearly so' (ibid., 105–6).
[40] Ibid., 105.
[41] Among the theories cited by Laudan are the effluvial theory of static electricity, the phlogiston theory, the caloric theory of heat, and the ether theory of light propagation (ibid., 113–23).

present context is the change of emphasis presented in *Science and Values*. Laudan's earlier emphasis on an inviolable procedure for assessing rational reconstructions of scientific progress has given way to a Reticulational Model of evaluative practice within science. His criticism of epistemic realism is based on the presumed correctness of the model, but the model itself does not stipulate what counts as progress in science. The Reticulational Model is basically a descriptive generalization about the way in which justificatory conflicts have been adjudicated within science. In so far as the model receives its warrant from the course of evaluative practice in science, it is subject to revision with the accumulation of experience. What counts as 'justification', 'constraint', 'realization', and 'harmony' may change over time.

8

Holton on Thematic Analysis

IT is in the case studies of Gerald Holton that the full potential of descriptive *PS* is achieved. Holton has uncovered methodological and valuational assumptions that underlie scientific practice within various episodes, and has traced continuities in evaluative practice over time.

What has emerged from these enquiries is a theoretical model which emphasizes the importance of 'thematic principles' within *HS*. Thematic principles express the basic commitments of scientists to preferred modes of enquiry and explanation. They include methodological prescriptions, evaluative standards, and high-level substantive hypotheses. Holton has suggested that a small number of these basic presuppositions has sufficed for the organization of science.[1]

Particularly important is the 'Ionian Enchantment', the ideal of a unified interpretation of all natural phenomena under a small number of fundamental laws.[2] Other themata identified by Holton include methodological standards such as 'seek within natural phenomena quantities that are conserved, maximized, or minimized', and 'seek to explain macroscopic irregularities by reference to underlying causal regularities'; evaluative standards such as parsimony, symmetry, and incorporation; and high-level hypotheses such as the constancy of the velocity of light, the discreteness of electric charge, and the quantization of energy.[3]

Holton observed that in some cases themata exist as members of a dialectically related pair of presuppositions. Examples include

---

[1] Gerald Holton, 'Thematic Presuppositions and the Direction of Scientific Advance' in *Scientific Explanation*, ed. A. F. Heath (Oxford: Clarendon Press, 1981), 16; *Thematic Origins of Scientific Thought: Kepler to Einstein* (Cambridge: Harvard University Press, 1973), 29.

[2] Holton, 'Thematic Presuppositions and the Direction of Scientific Advance', 17–23.

[3] Holton, *Thematic Origins of Scientific Thought*, 28–9; *The Scientific Imagination: Case Studies* (Cambridge: Cambridge University Press, 1978), 6–22; 'Do Scientists Need a Philosophy?' *Times Literary Supplement*, 2 Nov. 1984, 1235.

rival commitments to atomistic and plenistic theories, and rival preferences for mechanical and mathematical models.

Holton emphasized that thematic principles are neither empirical claims subject to straightforward experimental testing nor analytical truths devoid of empirical content. However, he cautioned against conferring upon these principles the status of Jungian archetypes or Kantian regulative principles of the reflective judgement. Thematic principles are subject to change within *HS*. Certain once-prevalent principles have subsequently been discarded. Among these are the principles of conservation of mass and conservation of parity, the assumption that there is a correspondence of structure at macrocosmic and microcosmic levels, and the explanatory requirement that a teleological argument-form be instantiated.[4] Holton has also shown that new thematic principles are formulated from time to time during the development of science. Bohr's Principle of Complementarity is an important example.

Holton issued no prescriptive recommendations on behalf of selected thematic principles. He did not claim, for instance, that scientists ought search for extremum principles, accept the constancy of the velocity of light, or evaluate competing theories by appeal to a criterion of simplicity. Indeed, he made no claims about how scientists ought proceed. His *PS* is an uncovering of principles that have guided research and warranted justificatory arguments within science.

Holton has concluded from his studies of important scientific episodes that science develops within a three-dimensional matrix, the axes of which represent empirical content, analytical content, and thematic content. He put forward this three-dimensional model in order to overcome the deficiencies of two-dimensional models which focus exclusively on the x–y plane on which empirical findings and analytical claims intersect.

Holton conceded that, for the most part, discussions within 'public science' take place on the x–y plane.[5] 'Public science' is that aspect of scientific activity in which conjectures and results are communicated within (and sometimes from) the community of scientists. He noted, however, that scientists themselves, in the creation of 'private science' in the 'nascent moment', invoke—

[4] Holton, *Thematic Origins of Scientific Thought*, 25.
[5] Ibid., 24.

consciously or otherwise—thematic principles that transcend the x–y plane.[6] Indeed, Holton maintained that, without reference to the thematic dimension, important episodes within *HS* seem at best puzzling and at worst irrational. Particularly good illustrations of this point are Einstein's work on Relativity Theory and Millikan's research on the charge of the electron.

Einstein refused to consider decisive Walter Kaufmann's 1906 experimental 'disproof' of Special Relativity Theory. Instead he stressed the economy of representation and breadth of scope of his theory and the *ad hoc* character of rival theories of electron motion. Only a decade later was Kaufmann's experimental result shown to be incorrect. Holton maintained that Einstein's 'willingness to suspend disbelief' in the face of recalcitrant data was a manifestation of Einstein's commitment to thematic principles which include:

primacy of formal (rather than materialistic or mechanistic) explanation; unity or unification; cosmological scale in the applicability of laws; logical parsimony and necessity; symmetry (as long as possible); simplicity; causality (in essentially the Newtonian sense); completeness and exhaustiveness; continuum; and of course constancy and invariance.[7]

R. A. Millikan, in the course of a dispute with Felix Ehrenhaft about the existence of a minimum unit of electric charge, selected for publication only those trials in his oil-drop experiment that supported the minimum-charge hypothesis. Holton noted that Millikan allowed his commitment to the hypothesis to determine which trials were 'genuine' and which trials had 'failed' (for instance, by focusing on a suspected dust particle rather than an oil drop). Holton's conclusion from these and other episodes was that 'any discussion of the advance of science that does not recognize the role of the suspension of disbelief at crucial points is not true to the activity'.[8]

Holton's comments on the achievements of Einstein and Millikan, as well as those of Kepler,[9] Newton,[10] Mach,[11] Bohr,[12] and Stephen Weinberg,[13] provide support for the three-dimensional

---

[6] Ibid., 17–20.
[7] Holton, 'Thematic Presupposition and the Direction of Scientific Advance', 15. [8] Ibid., 11.
[9] Holton, *Thematic Origins of Scientific Thought*, 69–90.
[10] Ibid., 49–53 [11] Ibid., 219–59, [12] Ibid., 115–61.
[13] Holton, *The Scientific Imagination*, 11–20.

model of science. He has shown that the basic categories of the model are operative in actual cases drawn from *HS*.

Holton's three-dimensional model, unlike Whewell's Tributary–River Model,[14] is not a model of scientific progress. It is, rather, a framework for the interpretation of the activities of scientists, whether or not these activities count as 'progressive'. As such, the three-dimensional framework functions within Holton's interpretation of science in much the same way that the fact–idea polarity functions within Whewell's interpretation of science.

Alternative frameworks for historical reconstruction may be compared. Holton argued that his three-dimensional model is superior to two-dimensional models because thematic considerations are involved in the analysis of problems raised by the following questions:

(1) What is constant in the ever-shifting theory and practice of science—what makes it one continuing enterprise, despite the apparently radical changes of detail and focus of attention?

(2) Why do scientists, at enormous risks, hold on to a model of explanation, or to some 'sacred' principle, when it is in fact being contradicted by current experimental evidence? [and]

(3) Why do scientists . . . with good access to the same information often come to hold so fundamentally different models of explanation?[15]

Holton insisted that no satisfactory answers to these questions are available solely upon consideration of changes recorded upon the x–y plane.

The first question poses the problem of the appropriate demarcation of science. Holton's own answer to the question is that the identity-through-time of the scientific enterprise derives from the set of thematic principles shared by participants. He declared that

on this model we can understand why scientists need not hold substantially the same set of beliefs, either in order to communicate meaningfully with one another in agreement or disagreement, or in order to contribute to cumulative improvement of the state of science. Their beliefs have considerable fine structure; and within that structure there is, on the one hand, generally sufficient stabilizing thematic overlap and agreement, and

---

[14] See Chapter 6, pp. 78–83.
[15] Ibid., 7.

on the other hand sufficient warrant for intellectual freedom that can express itself in thematic disagreements.[16]

Thus although descriptions of the x–y plane at different times may differ greatly, the relation of this plane to the z-axis provides an underlying continuity. Given this continuity of commitment to thematic principles, it is neither necessary nor helpful to superimpose upon the course of science such concepts as 'revolution', '*Gestalt-shift*', and 'conversion'.[17]

Although Holton maintained that it is commitment to thematic principles that is responsible for the continuity of science, he did not accord inviolable status to any particular thematic principle. On Holton's model, themata are subject to modification or deletion. Moreover, a given thematic principle may be related to an antithetical principle, such that the point of balance between them changes over time. It is the task of the philosopher of science to record these changing relationships within the three-dimensional fabric of science.

Holton's three-dimensional model also provides answers to questions (2) and (3) above. It is commitments to thematic principles which, in certain cases, account for scientists' seeming inability to see the force of negative experimental evidence. In addition, scientists' commitments to divergent thematic principles may be responsible for their pursuit of different models of explanation.

Critics of a descriptive approach to *PS* may concede the fact that thematic analysis provides answers to questions (2) and (3), but object that to answer such questions is to practise history or psychology. On this view, the philosopher of science should seek to answer questions about the logical and epistemological status of 'public science' as recorded in the appropriate literature. The 'private science' of the 'nascent moment' is not his concern *qua* philosopher.

Consider the recent interest in rational reconstructions of scientific progress. Such reconstructions contrast that portion of *HS* that conforms to selected evaluative criteria ('internal *HS*') with that portion of *HS* that does not so conform ('external *HS*').

---

[16] Holton, 'Thematic Presuppositions and the Direction of Scientific Advance', 24. See also *The Scientific Imagination*, 11.

[17] Holton, 'Thematic Presuppositions and the Direction of Scientific Advance', 25.

From the standpoint of defenders of a prescriptive *PS*, the fact that scientists have contributed to 'external *HS*' is regrettable, but analysis of the reasons why they failed to implement 'proper' standards is a task for historians and psychologists.

Holton is completely opposed to such an approach to the interpretation of science. It is his position that what counts as 'internal *HS*' ought to be determined by reference to the three-dimensional model, and not by reference to some a priori criterion of scientific progress.

Suppose Holton is correct that certain instances of methodological behaviour that appear perverse within two-dimensional frameworks are accommodated by the three-dimensional model. Why should this count in favour of the three-dimensional model? It is open to Holton to stipulate that the three-dimensional framework is a fundamental starting-point for the interpretation of science, a starting-point not subject to justification by appeal to more basic considerations. An alternative would be to argue that applications of the three-dimensional model achieve an otherwise unavailable accommodation between theoretical interpretations of science and our preanalytic intuitions about scientific progress. Instances of prima facie perverse methodological behaviour are now seen to be manifestations of commitments to thematic principles. To defend this position would be to accept the achievement of reflective equilibrium[18] as the standard by which conceptual frameworks for the interpretation of science are to be evaluated.

Holton's prescriptive recommendation of the three-dimensional framework does not compromise the programme of descriptive *PS*. The three-dimensional framework is not a blueprint for scientific progress. Moreover, as emphasized above, Holton has not sought to justify individual evaluative standards for science. Nor has he sought to justify a three-dimensional model based on a particular set of thematic principles. He has not argued, for instance, that a model that emphasizes the thematic principle of discreteness is superior to a model that emphasizes the thematic principle of continuity. His justificatory argument establishes only that it is important not to overlook the role of thematic principles in the development of science.

I suspect that the principle that thematic analysis is essential to

---

[18] See Chapter 6, pp. 115–16.

an interpretation of science functions as an inviolable principle within Holton's *PS*. It is not inconsistent with the conclusion of Chapter 5 that there exist a descriptive *PS* within which certain principles are held to be inviolable. The argument of Chapter 5, if correct, establishes only that every prescriptive *PS* contains an inviolable principle.

Given the inviolable principle that thematic analysis is essential to *PS*, the remainder of Holton's *PS* conforms to Shapere's vision of a *PS* devoid of inviolable principles. The price paid for this achievement is abandonment of Shapere's requirement that the evaluative criteria of *PS* retain prescriptive status.

History of Science and Descriptive Philosophy of Science

UNQUALIFIED Historicism and Unqualified Logicism are the extreme positions on the role of *HS* in the selection and justification of prescriptive philosophies of science. These positions, and positions intermediate between the extremes, have been examined in earlier chapters.

The role of *HS* in the selection and justification of descriptive philosophies of science remains to be investigated. I take it to be non-controversial that *HS* is a discipline whose aim is to reconstruct the course of scientific development. The historian of science, *qua* historian, does not issue pronouncements about how science ought be practised. Neither does the philosopher of science who accepts a descriptive approach to his discipline. Given that the subject matter of *HS* is the totality of scientific practice and that the evaluative judgements of scientists are judgements integral to the practice of science, it might seem that descriptive *PS* is subsumed under *HS*. If this conclusion were correct, then the philosopher of science would be a historian with a special interest in past evaluative practice. A *PS* subsumed under *HS* would lack autonomy. The appropriate investigative approach and disciplinary standards would be those of history. However, it does not follow from the fact that the first-order subject matter of *HS* includes that of *PS* that *PS* is subsumed under *HS*. Certainly *PS* is not subsumed under *HS* as the class of triangles is subsumed under the class of polygons. Every predicate properly predicated of polygons in general is properly predicated of triangles. But *PS* is not a subclass of *HS* in this sense. Kuhn is correct to insist that philosopher and historian approach a common first-order subject matter with different interests. According to Kuhn, the historian seeks to create a narrative which has explanatory force, a narrative within which the facts fall into a recognizable pattern.[1]

---

[1] Thomas S. Kuhn, 'The Relations Between the History and the Philosophy of Science' in *The Essential Tension* (Chicago: University of Chicago Press, 1977), 17–18.

This description of the historian's intent is cogent, even if Kuhn's specific Puzzle-Solving Model of historical explanation is not.[2] The philosopher, by contrast, is not interested in narration *per se*. His interest is in uncovering the standards that underlie individual evaluative judgements. It is the nature of a standard that it applies to a multiplicity of cases, actual and/or potential. According to Kuhn, this interest in that which is general sets the philosopher apart from the historian. Kuhn is correct to insist that *PS*, even in its descriptive version, remains a normative discipline.

The difference of interests of which Kuhn speaks is real. However, the distinction between descriptive *PS* and *HS* is not a sharp one. The activities of the descriptivist philosopher of science and the historian of science overlap. The descriptivist philosopher of science must examine individual evaluative decisions of scientists in order to identify the standards that underlie this practice. And the historian of science must be sensitive to the pervasive influence of evaluative standards in order to create effective explanatory narratives.

I shall continue to discuss the programme of 'descriptive *PS*', while conceding that, on a broad view of the scope of *HS*, the tasks assigned the descriptivist philosopher of science are tasks integral to historical reconstruction. In the last analysis, it does not matter very much whether the search for evaluative standards is labelled 'philosophical' or 'historical'. To make this concession is to acknowledge that there is a measure of vagueness to questions about the role of *HS* in the justification of descriptive claims about the efficacy of evaluative standards in science.

Defenders of a prescriptive version of *PS* may object to the very idea of a descriptive *PS*. One likely prescriptive argument is that the proper focus of *PS* is the context of justification and that a *PS* without prescriptive intent would be exclusively a study of the context of scientific discovery. According to this objection, a purely decriptive *PS* is not a 'philosophy of science' at all.

This objection is not on target. Descriptive *PS* does contain justificatory arguments. However, the aim of such arguments is not to justify proposed evaluative standards, but rather to justify claims about actual evaluative practice. The descriptivist philosopher of science argues, not that evaluative standard *S* ought be

[2] See Chapter 2.

implemented, but that S is a standard, commitment to which has been important within the context of specific scientific episodes. The appropriate justification of such a claim is to show that S indeed was effective in the manner claimed.

Suppose it is claimed that evaluative standard S was important in historical episode E. What sorts of justificatory arguments are available to support the claim?

If the scientists whose activities constitute E claim that they applied S in the course of their research, then this claim provides support for the importance of S in E. However, the explicit testimony of the appropriate scientists is not a sufficient condition of the correctness of the thesis about the importance of S. Scientists have been, and presumably may continue to be, deluded about their commitments to evaluative standards.

The more usual case is that the scientists involved in E make no claims about the application of evaluative standards. In such a case, a justificatory argument in support of the importance of S may invoke a subjunctive conditional claim of the form

> if S had been explicitly selected and applied by the participating scientists, then the evaluative decisions rendered would be consistent with the events which comprise E.

Since one cannot replay the evaluative situation under consideration, support for the subjunctive conditional claim is necessarily indirect. For instance, one might appeal to what happened in other evaluative situations in which it is believed that S was applied. This is a point at which evidence from *HS* is relevant to the appraisal of descriptive interpretations of evaluative practice.

In addition, one descriptive interpretation of evaluative practice may be judged superior to a second because it better accommodates agreed-upon 'benchmark facts' about disputed episodes. 'Benchmark facts' are non-controversial details about events. For example, all interpreters may be in agreement that a scientist performed a certain sequence of actions—for example Jones constructed a detection circuit, threw a switch, recorded meter readings, made calculations, and published a paper in which he claimed that these calculations refuted a hypothesis championed by Smith. Of course, any of these facts may be challenged. But it may be the case that no disputes arise at this level of interpretation, in which case analyses inconsistent with these facts may be disqualified. Since benchmark facts are recorded in histories of science, this is another point at which an appeal to *HS* is relevant to claims about evaluative standards.

The interpretations of descriptive *PS*, like the interpretations of *HS*, reflect judgements of importance about the first-order subject matter. A given set of judgements of importance may be challenged from the perspective of a rival interpretation. The first interpretation may be criticized because it fails to take into account factors judged important on the rival view. Or it may be criticized because the factors which it does take into account are inappropriately weighted. Challenges of these kinds are effective only if supported by the historical record. Support is forthcoming if the types of judgements emphasized on the rival view can be shown to have been fruitful in other contexts.

Waismann's description of how a philosopher 'builds a case' for a position is a good description of the justificatory strategy of the descriptivist philosopher of science. According to Waismann, the philosopher

makes you see all the weaknesses, disadvantages, shortcomings of a position; he brings to light inconsistencies in it or points out how unnatural some of the ideas underlying the whole theory are by pushing them to their farthest consequences. . . . On the other hand, he offers you a new way of looking at things not exposed to these objections.[3]

One may pursue a justificatory strategy of this type without a commitment to specific inviolable principles.[4] Needless to say, there is room for honest disagreement among practitioners about the soundness of particular justificatory arguments.

The descriptivist philosopher of science who does succeed in 'building a case' for the importance of particular evaluative standards in specific episodes has made a contribution to our understanding of past science. Holton, for instance, has shown that there has been a large measure of continuity within the thematic dimension of evaluative practice.

The programme of descriptive *PS* is worth pursuing for its own sake, regardless of whether or not prescriptive recommendations

[3] Friedrich Waismann, 'How I See Philosophy' in *Logical Positivism*, ed. A. J. Ayer (New York: Free Press, 1959), 372–3.
[4] The descriptivist philosopher of science presumably does accept the 'principle' that one ought respond to demands for a justification of one's interpretations.

are extrapolated from, or 'justified' upon appeal to, its findings.[5] Feyerabend's counsel to 'return to the sources' is good advice.[6]

[5] Difficulties which beset the programme of prescriptive *PS* are discussed in Chapters 5 and 6 above.
[6] P. K. Feyerabend, 'Philosophy of Science: A Subject with a Great Past' in *Historical and Philosophical Perspectives of Science*, ed. R. Stuewer (Minneapolis: University of Minnesota Press, 1970), 183.

Index of Proper Names

Index of Subjects